# APPLIED SOLAR ENERGY

Th s b... be...

# APPLIED SOLAR ENERGY

A guide to the design, installation and maintenance
of heating and hot water services

David Kut
*and*
Gerard Hare

south essex college
FURTHER & HIGHER EDUCATION
**Skills** | Education | Careers

South Essex College
Thurrock Campus

**THE ARCHITECTURAL PRESS: LONDON**
**HALSTED PRESS DIVISION**
**JOHN WILEY & SONS, NEW YORK**

First published in 1979 by The Architectural Press Limited:
London

© 1979 David Kut and Gerard Hare

**British Library Cataloguing in Publication Data**

Kut, David
    Applied solar energy.
    1. Solar heating
    I. Title II. Hare, Gerard
    697'.78         TH7413

ISBN: 0 85139 046 3

Published in the USA by Halsted Press
a Division of John Wiley & Sons, Inc, New York

**Library of Congress Cataloging in Publication Data**

Kut, David.
    Applied solar energy.
    "A Halsted Press book."
    Includes bibliographical references and index.
    1. Solar heating—Handbooks, manuals, etc. 2. Solar
water heaters—Handbooks, manuals, etc. I. Hare, Gerard,
joint author. II. Title.
TH7413.K87  1979        697'.78        79-15879

ISBN 0-470-26801-8

Filmset in 11/12 point Plantin
Printed and bound in Great Britain
by W & J Mackay Limited, Chatham

# CONTENTS

# PREAMBLE

This book is intended to guide the installer of solar heating systems and to enlighten the layman who seeks the unadorned facts about this subject.

The text and the accompanying illustrations are mainly concerned with the application of solar energy to hot water supply, space heating and swimming pool systems, but reference is also made to other uses of solar energy which are currently undergoing development or research.

The practical aspects of selecting, assembling, maintaining and repairing a variety of installations and components are given prominence; in particular, Gerard Hare has fed into this book the results of much practical experience in the design, manufacture and fitting of solar collectors.

For the convenience of the reader, the text employs both metric and imperial units.

The authors hope that this book will contribute materially towards the better understanding of solar energy systems and to their successful application.

David Kut and Gerard Hare

# ACKNOWLEDGEMENTS

The authors acknowledge and appreciate the assistance obtained from the following persons and firms in respect of information relating to their products or research which have been incorporated in this book: Ralph Hopkinson and Newton Watson; Daniels Engineering Ltd.; Lesney Products U.K. Operations; Mr. M. Pickett and Mr. J. Stein; Mr. E. Tear; Dr. J. C. McVeigh; W. S. Atkins & Partners; Electra (Israel) Ltd.; Wolfson Foundation Report May 1977; Sealed Motor Construction Co. Ltd.; Satchwell Control Systems Ltd.; Electra Air Conditioning Services Ltd.; T. H. Thorpe & Partners, architects in respect of the solar heated swimming pool at the School of St. Mary & St. Anne, Abbots Bromley.

Finally, we acknowledge the able assistance provided by Mrs. Jean Hollis (David Kut & Partners), Mrs. Eileen Sheppard (Gerard Hare (Engineers) Ltd.) and the cheerful sketch of the cat provided by Debbie Kut.

# 1

# INTRODUCTION

Nature, left to itself, constantly degrades humanity's energy input into heat. In the exploitation of solar energy, this process is reversed; the heat input is converted into useful work. In its perverse way, nature decrees that the great bounty of solar energy is unevenly distributed: there is a vast outpouring to the great deserts and arid zones of the earth, whilst most of the developed regions suffer shortage, see Figs. 1.1 and 1.2. In the UK, solar input is moderate during the summer and much reduced during the winter and autumn, when there is the greatest demand for heat to warm homes and places of work.

The nearest star is the sun. It is of average size with a diameter of about 1,376,000 km (860,000 miles); it rotates on its axis about once a month and, being a mass of incandescent gas, a gigantic nuclear furnace where hydrogen is built into helium, its density is slightly less than 1½ times that of water. The process of energy change has been going on for billions of years and will continue for billions more until the sun becomes a 'red giant' – exhausted of most of its major energy potential. Meanwhile, some four million tons of the sun's matter will continue to be changed into energy every second. The sun's dazzling surface, the photosphere, is speckled with dark sunspots and bright patches. Great prominences or streamers of hot gas shoot out and rain down onto the surface of the sun, moving at different speeds. Around the sun is a permanent halo or corona of rising and falling gas, so intense that observation of this brilliant spectacle with the naked eye is unsafe. Only about one two-billionth of this energy from the sun reaches the surface of the earth, see Fig. 1.3.

1

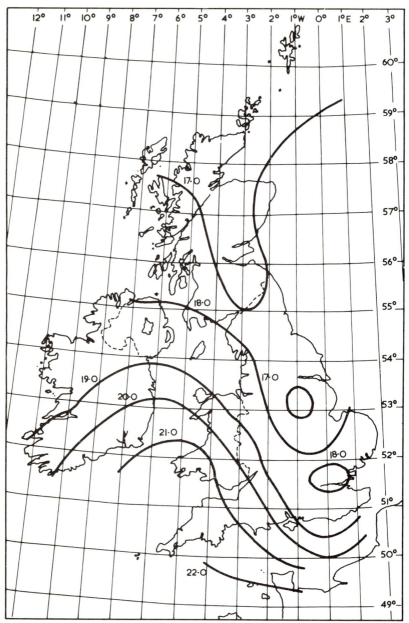

MID SUMMER (JUNE) GLOBAL SOLAR RADIATION
AVERAGE DAILY TOTALS IN MJ/m²

1.1    Global solar radiation – mid-summer

Readers of advertisements and of technical papers may be forgiven for thinking that man's endeavours towards harnessing the sun's energy were born in the agitation and wake of the recent predictions of approaching fuel shortages and ever-escalating prices. Such views ignore the long record of progressive achievements in this field by many distinguished scientists from Joseph Priestley and Lavoisier (1774) to our present-day researches and developments. We, of course, have the more sophisticated tools and suitable materials produced by advanced technology.

'Conventional' heat energy derives from the sun. Photosynthesis converts the incident solar energy into plant growth; the vegetation rots away, and the ever greater pressures and temperatures exerted upon this composted, re-cycled material during the passage of thousands upon thousands of years, processes this gradually into the solid, liquid and gaseous fuels which warm our homes, heat our food, propel our transport and progress the march of industrial sophistication.

Every two days, the sun showers the earth with energy equivalent to all the known reserves of stored fuel on earth. Mankind owes its present relatively comfortable standard of life and the expectations of further advancement to this hitherto abundant and readily accessible store of fuel.

Crude oil and natural gas currently contribute two-thirds of the world's consumption of energy and, in spite of rapidly escalating costs, are expected to increase their share further. This onrush in demand has created much concern for the safety of long-range energy supplies in areas which depend to a major extent on oil and gas. Apart from the enormous out-flow of funds from the oil-consuming countries, a more basic concern is the capability of the major exporters of crude oil to achieve a continuing production target of roughly doubling output every ten years to meet the required oil demand if present trends continue.

Against this background of the depletion of our conventional fuels, engineers and scientists are now turning their eyes and ingenuity towards the more direct use of the truly fantastic outpouring of energy from the sun for our everyday existence and essentials. Equipment designed for the conversion of the energy of the sun to heat or power utilises the solar radiation for this purpose. The main factors which influence the quantities of radiation received by a horizontal surface at ground level are the position of the sun in altitude and azimuth (see chapter 3). Most important meteorological stations record sun radiation in terms of the total or

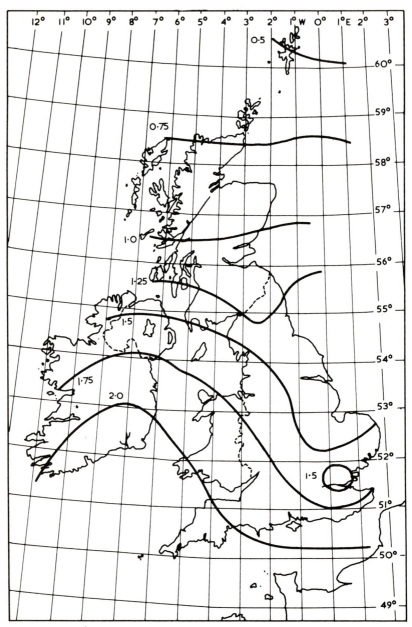

MID WINTER (DEC) GLOBAL SOLAR RADIATION
**AVERAGE DAILY TOTALS IN MJ/m²**

1.2    Global solar radiation – mid-winter

global radiation received by a horizontal surface as measured by a thermo-pile type radiometer. The alternative practice of estimating solar radiation in terms of sunshine hours has been found neither satisfactory nor scientifically accurate.

The global solar radiation comprises three basic components:
    Direct radiation
    Diffuse radiation
    Reflected radiation from the ground.

For the purposes of computations, the emitter of solar radiation – the sun – may be assumed to be a point source of direct radiation, and the magnitude of such radiation in a cloudless sky can be calculated from the geometry of the solar position angles.

The diffuse radiation is received from the whole sky hemisphere and is conveniently assumed to be of uniform intensity; this assumption is clearly not strictly accurate, but is adequate for most purposes.

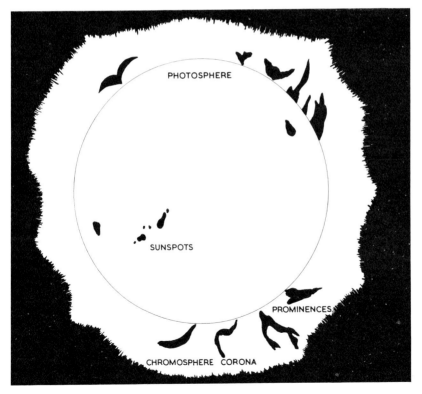

1.3   The sun – source of solar energy

Under some conditions, there may be solar radiation reflected from the ground onto inclined surfaces; such radiation is difficult to assess and is more commonly ignored in most practical solar energy calculations.

It is only worthwhile installing solar radiation-based energy equipment in areas where one can be reasonably assured of an adequate supply of such radiation. Table 1.1 indicates the average levels of solar radiation received on a horizontal surface in various areas of the earth. The use of the term 'average' is stressed, as there can be very wide local variation in the reception of solar energy by reason of geographical and weather-pattern; this is especially relevant to solar conditions in the UK.

**Table 1.1   Total or global radiation on a horizontal surface**

| Country | Daily average over the year | | Annual total GJ/m$^2$ |
|---|---|---|---|
| | Wh/m$^2$ | Btu/ft$^2$ | |
| Northern Europe | 240–280 | 760–890 | 3.1–3.8 |
| Mediterranean countries | 430–600 | 1360–1900 | 5.6–8.0 |
| Sahara Desert | 635 | 2010 | 8.3 |
| Central Africa | 440–630 | 1400–1990 | 5.8–8.3 |
| South Africa, desert | 700 | 2210 | 9.1 |
| central high veld | 570 | 1810 | 7.5 |
| coastal regions | 465 | 1475 | 6.1 |
| India and Pakistan | 520–630 | 1660–1990 | 6.8–8.3 |
| Australia, Darwin | 565 | 1790 | 7.4 |
| Alice Springs | 610 | 1930 | 8.0 |
| Melbourne | 410 | 1310 | 5.4 |
| Japan | 325–420 | 1030–1330 | 4.2–5.6 |
| USA Northern States | 385 | 1220 | 5.0 |
| UK | 267 | 847 | 3.5 |
| USA Southern States | 580 | 1840 | 5.6 |
| Canada, North | 230 | 740 | 3.1 |
| South | 355 | 1125 | 4.7 |

Table 1.1 shows that areas with considerable scope for solar energy are the Mediterranean countries, South-East Africa, India and Pakistan, Australia and the Southern States of the USA. The solar energy falling each year on each acre in India or Egypt equals that obtained by burning efficiently more than 1,000 tons of coal

and some of these countries already use relatively cheap commercially available domestic type solar water heaters. Japan is not in the top league of solar energy recipients, but by the year 1970 over 2.5 million commercial solar water heaters had been sold for use in that country.

The solar radiation in the UK incident in an average year on each square metre of south-facing roof, averages about 3.5 GJ (3.1 therm/ft²) per annum; of this some two-thirds is received during the summer.

The typical household in the UK has net annual requirement for space heating of about 52 GJ (491 therm) and for hot water supply of about 8 GJ (173 therm). During the summer season, there is a relative abundance of sunshine and this can be utilised to heat or pre-heat domestic hot water, the fewer sunny days during the remainder of the year giving an added bonus. The space heating requirement is large, relative to the average annual solar energy availability, and must be met in a season of minimum sunshine.

Given the present techniques of solar energy collection and storage, space heating by solar energy cannot be even remotely economically viable in the UK. However, such a negative conclusion does not apply in the many countries in which a hot sunny day is usually followed by a chilly or cold evening and night; adequate solar heat can be collected and stored to supply domestic hot water, as well as space heating, during the day to meet both these needs over each period of twenty-four hours.

*References:*

R. G. Courtney, *An Appraisal of Solar Water Heating in the UK*, Building Research Establishment, January 1976

R. G. Courtney, *Solar Energy Utilisation in the UK: Current Research and Future Prospects*, Building Research Establishment, October 1976.

Farrington Daniels, *Direct Use of the Sun's Energy*, Ballantine, New York, 1975

D. Kut, Solar Energy Applied to Comfort and Heat Services, *Heating and Air Conditioning Journal*, p.6, October 1974

# 2

# DEFINITIONS

A householder, faced with a proposal or estimate for a solar energy collection system, will be better equipped to understand it if he has a knowledge of the meaning of the basic technical terms employed therein. Some salesmen tend to use terms loosely, without a full understanding of their correct meaning. This chapter first lists and then defines the terms most commonly used in the solar energy field.

Absorptance
Absorption
Absorption
co-efficient
Air conditioning
system
Air mass
Airvent
Algae
Altitude
Ambient
Anti-freeze
Anti-reflection
coating
Auxiliary energy
Array
Azimuth
Ball valve

Black body
Circulating pump
Clearness
Colour temperature
Control (devices)
Corrosion
Dezincification
Delta T
(temperature)
Dezincification
Diffuse radiation
Direct radiation
Efficiency
Electric wiring
system
Electrolytic action
Emittance
Equinox

Evacuated tubular
collector see Solar
collectors
Extended finned
surface
Fan
Flat plate collector
Flux
Frost Damage
Frost protection
Greenhouse effect
Heat exchanger
Heat store
Heat transfer fluid
Hour angle
Incident radiation
see Radiation
Inclined surface

**Absorptance**     The ratio of the radiant energy absorbed by a plane surface to the radiant energy incident upon that surface. (See Fig. 2.1.)

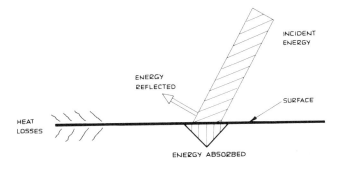

2.1   Absorptance

**Absorption**     The process in which radiation is converted within a material into excitation energy. Most of the radiant energy falling (or incident) upon a matt black surface is absorbed

– usually not less than 80% and with specially treated or selective surface as much as 98%. The process of thermal conversion is complex, involving acceleration of electrons, multiple collisions, photon absorption and scattering. Radiant energy of all wavelengths is thus degraded into heat. *See* 'selective surface'.

**Absorption coefficient**
A measure of the absorbing strength of a material for radiant energy per unit length. Usually expressed more simply as a decimal (0.8 or 0.98) and denoted by letter $\alpha$.

**Air conditioning system**
Comprehensive air conditioning correctly distributes treated air into a space and achieves pre-determined conditions in that space of air temperature, moisture content, filtration and air flow.

**Air mass**
The length of the path through the earth's atmosphere traversed by direct solar radiation, expressed as a multiple of the path length with the sun at zenith.

**Air vent**
A device for removing air from a liquid system, either manually or automatically.

**Algae**
Unicellular or filamentous plants that are usually fast-growing and live in fresh or sea water.
Algae formation within solar collectors is most undesirable.

**Altitude**
The angle which the rays of the sun make with the horizontal plane at a given point.

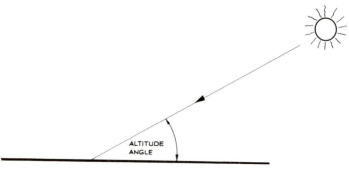

ALTITUDE ANGLE

2.2    Altitude

| | |
|---|---|
| **Ambient** | Surrounding (temperature). |
| **Anti-freeze** | Prevents freezing, usually in water, by various additives or by heating trace elements keeping fluid temperature above freezing point $-0°C$ ($32°F$). |
| **Anti-reflection coating (abbreviated as AR)** | A coating applied to a surface (bloomed) to increase the amount of light penetration. May be applied to the surface of solar cells or to glass/plastic covers of solar collectors. |
| **Auxiliary energy** | The use of an alternative energy source, such as solar, lessens the amount of primary energy (from gas, electricity, etc.) otherwise required. The primary energy provision required is referred to as the auxiliary or back-up need. |
| **Array** | *See* 'solar array'. |
| **Azimuth** | The angle between the horizontal component of the rays of the sun and the true south. (See Fig. 2.3.) |
| | Azimuth angle is usually measured in degrees east (morning) and degrees west (afternoon) of south. |
| **Ball valve** | In most open-type water cisterns there is a requirement for automatic water replen- |

2.3   Azimuth

ishment or make-up. A ball valve usually comprises a metal or plastic float linked by a lever-arm to a slide valve mechanism of a mains water valve. (See Fig. 2.4.)

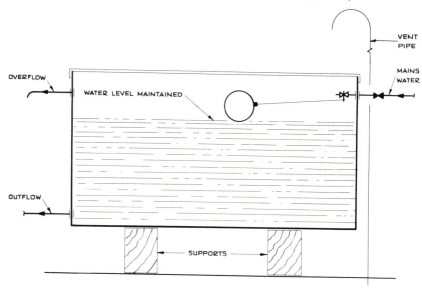

2.4   Tank with ball (float) valve

**Black body**

This describes an ideal substance which absorbs all the radiation falling upon it and emits nothing. An alternative definition: a body which emits the maximum possible radiation ie, its emissivity is 1.0.

**Circulating pump**

A device (usually electrically driven) used to move or circulate a liquid (most commonly water) through a pipe to transfer heat from, say, a solar collector to a heat store. The human heart is a good example of a circulating pump.

**Clearness**

The atmosphere is not usually as clear at sea-level as at, say 1 mile (1.6 km) above sea-level; thus the radiation level or insolation is not as valuable. This phenomenon is defined by 'clearness' numbers. In parts of India and in the Arizona Desert, clearness numbers approach 1.0.

**Colour temperature**  The temperature, in degrees K, at which the shape of the spectral irradiance of a full black body radiator most closely matches the shape of the spectral irradiance curve of the radiant source under consideration. The colour temperature for global solar radiation over the wave band 300 nm to 780 nm at the earth's surface in the United Kingdom is typically around 5,700°K, with substantial variations.

**Control (devices)**  The successful collection of heat requires control to avoid overheating, boiling or frost damage and to achieve diversion into and out of the heat store. Such control devices, electrically or thermo-mechanically actuated, are applied to solar collection systems.

**Corrosion**  Corrosion in solar collection systems usually causes 'pitting' or chemical attack on equipment surfaces. Consequences of corrosion may be serious. See chapter 5.

**Declination**  The angle between the plane of the earth's orbit and the equatorial plane; or the angular distance of the sun from the equator.

**Delta T (temperature)**  The difference in temperature between the mean fluid temperature of a solar collector and the ambient, or surrounding, air temperature.

**Dezincification**  A form of corrosion which results in the preferential removal of zinc from the duplex brass microstructure; it converts brass to porous copper and can lead to blockages by corrosion products and to loss of strength.

**Diffuse radiation**  *See* 'radiation'.

**Direct radiation**  *See* 'radiation'.

**Efficiency**  The ratio of $\dfrac{\text{(energy) output}}{\text{(energy) input}}$

A measure of the effectiveness of any device. *See* 'solar collector (efficiency)'.

**Electrical wiring system**  Electric power required to drive a circulating pump and other powered devices, together with low-voltage control to switch this device off and on, necessitates an electrical wiring system; usually an extension from the household or building general electric supply.

**Electrolytic action**

The cause of deterioration or failure of dissimilar metals (most commonly copper and galvanised mild steel) in contact with each other, accelerated by the presence of a warm/hot fluid, eg, failure of a combined copper and galvanised steel pipe carrying hot water within a circulating loop.

**Emittance**

The ratio of the radiant energy emitted (in the absence of incident radiation) from a given plane surface at a given temperature, to the radiant energy that would be emitted by a perfect black body at the same temperature.

**Equinox**

The moment at which the sun apparently crosses the celestial equator – the point of intersection of the ecliptic and the celestial equator when the declination is zero.

**Evacuated tubular collector**

*See* 'solar collectors'.

**Extended finned surface**

By extending the area of metal surface attached to a tube conveying a fluid, the amount of heat transferred to the liquid is increased. Fins may be rectangular or circular, attached by a mechanical or soldered bond.

**Fan**

A device (usually electrically driven) to move air against resistance, for example to circulate air between residence and heat store for the purpose of transferring heat and ventilation air. (See Fig. 2.5.)

**Flat plate collector**

*See* 'solar collectors'.

**Flux**

A substance mixed with metal to promote fusion, as with silver-solder jointing of copper pipework.

**Frost damage**

Water freezes at 0°C (32°F). In so doing, it increases in volume (change of state –

2.5 Axial fan (cut-away view)

liquid to solid), thereby expanding the pipe or container. Frost occurs at temperatures below freezing point, but is not usually revealed until the pipe is 'thawed' or unfrozen.

**Frost protection**  There are several methods of preventing frost damage to pipes and solar collectors: by emptying, by electrical or other trace heating, by using a liquid which freezes at much lower temperatures and by efficient thermal insulation.

**Greenhouse effect**  The air temperature under a glass or transparent cover increases when subjected to heat radiation. This 'effect' is caused by the absorption of radiation by the surfaces under this transparent cover and by the ready absorption of radiation in the long-wave length, or infra-red, band being unable to re-radiate through the transparent cover.

**Heat exchanger**  A vessel or heat store incorporating a primary pipe coil or annulus; heat is transferred from the primary fluid to the cooler secondary fluid. (See Fig. 2.6.)

**Heat store**  A vessel, tank or solid material in which heat is accumulated over a period of time

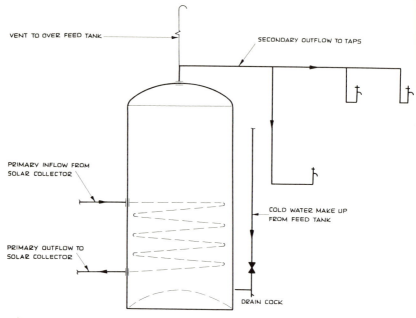

VENT TO OVER FEED TANK

SECONDARY OUTFLOW TO TAPS

PRIMARY INFLOW FROM
SOLAR COLLECTOR

COLD WATER MAKE UP
FROM FEED TANK

PRIMARY OUTFLOW TO
SOLAR COLLECTOR

DRAIN COCK

2.6   Heat exchanger

**Heat transfer fluid**

for subsequent use, when required.
Special fluids with a low freezing point are sometimes used, instead of water, to transfer heat from a solar collector tank to a heat store as a primary medium.

**Hour angle**

The angular distance of the sun from its position at noon.

$$h = \frac{360}{24} \times T$$

T is the number of hours of the sun time, either side of noon.

**Incident radiation**   *See* 'radiation'.

**Inclined surface**   A surface or solar collecting device tilted at an angle to the horizontal plane or to the observer's horizon.

**Infra-red**   Invisible long wavelength radiation (heat).

**Insolation**   The solar energy incident on a unit area of surface over a period of time; the time-integrated solar irradiance.

**Insulation**   Thermal wrapping or lagging applied to a

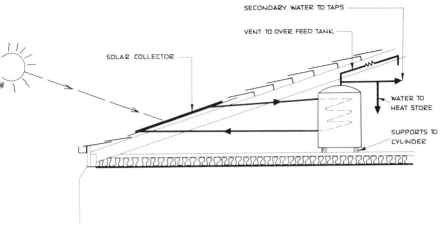

2.7    Solar collector – inclined surface

**Irradiance**

heat store or pipe to reduce heat loss or to protect against frost.

The radiant energy falling per unit area on a plane surface per unit time, normally stated in watts/metre$^2$ or Btu/ft$^2$.

**Latitude**

The latitude of a point on the earth's surface is its angular distance from the equator. For example, the latitudes of the following cities are:

London      51° 30′N
New York    40° 40′N
Sydney      33° 52′S
Delhi       28° 38′N

**Local apparent time (lat)**

System of astronomical time in which the sun always crosses the true north-south meridian at 12 noon. This system of time differs from local time according to longitude and time zone. The precise displacement also varies with the time of year.

**Longitude**

The angle which the terrestrial meridian through the geographic poles and a point on the earth's surface makes with a standard meridian (usually at Greenwich, UK).

**Net long wave radiation**

The net radiation, excluding the incoming and outgoing short wave radiation.

**Net present value analysis (NPV)**   A technique for analysing the economics of systems which have a long life expectation, involving discounting future economic benefits to their present value and using appropriately selected interest rates. The assumed interest rate is critical to the outcome of the analysis.

**Net radiation**   The difference between the total incoming and total outgoing radiation. Incoming radiation is made up of (a) short wave from sun and sky and (b) thermal, mainly from atmospheric layers close to the ground; outgoing radiation from a surface comprises (a) reflected short wave component depending on incident irradiance and reflectance of the surface, and (b) long wave thermal radiation component depending on ground surface temperature and long wave emittance.

**Passively heated solar buildings**   A building where solar input is stored and used internally without the aid of mechanical equipment, such as collector and pumps.

**Payback condition (simple)**   Economic stipulation which requires that the achieved life of a (solar) system shall exceed the initial cost of the (solar) system, divided by the annual value of the fuel saved by installing the system, ignoring all interest charges and inflation.

**pH value**   The potential of hydrogen. An index used to signify whether water is neutral, acidic or alkaline.

**Radiation**   The radiant energy falling on a plane per unit area integrated over a stated period (day, month, year, etc). Normally stated in megajoules/$m^2$ ($mj/m^2$) over the stated period.

**Radiation (diffuse)**   The scattered radiation falling on a plane of stated orientation over a stated period from the sky, and, in the case of an inclined surface, reflected from the ground as well.

**Radiation (direct)**   The radiation from the sun falling on a plane of stated orientation over a stated period received from a narrow solid angle centered on the sun's direction.

**Radiation (incident)** The combined diffuse and direct radiation components calculated proportionately to the fraction of sky hemisphere to which the plane is exposed and also calculated vectorially.

**Reflectance** The ratio of the radiant energy reflected from a surface, to the radiant energy incident upon that surface.

**Roll bond process** Bonding sheets of metal together by simultaneous rolling and heat treatment process.

**Selective surface** A surface which has high absorbance for incident radiation (wavelengths less than 1,000 nm) and high reflectance (low absorbance) in the infra-red range (wave lengths greater than 1,000 nm).

**Solar array** A number of individual solar collection devices arranged in a specific pattern (to collect solar energy effectively).

**Solar collector (concentrator)** A reflector system to increase sunlight intensity on a given area. (See Fig. 2.8.)

**Solar collector (concentration ratio)** The ratio of the heat flux within the image, to the actual heat flux received on earth, at normal incidence.

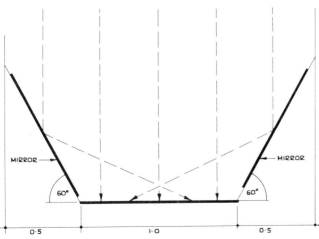

2.8    Solar concentrator

**Solar collector
(efficiency)**

An indication of relative performance or character, usually indicated on a graph incorporating important variables shown in the 'Hottel-Whillier-Bliss' equation:

$$Q = F(\alpha I - U\Delta T) - 2.1$$

in which

Q = useful heat collected
F = heat transfer effectiveness
$\alpha$ = transmittance absorptance product
I = incident solar radiation
U = heat loss coefficient
$\Delta T$ = temperature difference (collector mean temperature – ambient air temperature).

Collector efficiency varies from sunrise to sunset and especially when the collector is of the fixed, non-tracking type. Normal increase of Delta T also seriously affects the performance of some collector types.

**Solar collector
(evacuated tube)**

An alternative approach to reducing collector heat losses by employing a partial vacuum within transparent tubes arranged in parallel form over a reflector plate. Within each tube are usually one or more absorber tubes.

**Solar collector
(flat plate)**

Any non-focusing, flat-surfaced solar heat collecting device. (See Figs. 2.9 and 2.10.)

**Solar collector
(parabolic)**

A focusing type of solar collector, usually arranged in trough form, having a line-focus.

**Solar collector
(paraboloid)**

A focusing type of solar collector produced from the rotation of a parabola around its axis. The concentration rate will be the square of that for the parabola.

**Solar collector
(tracking)**

A mechanised solar collector arranged to follow or track the path of the sun and normalise the angle of incident radiation falling upon the collector surface.

**Solar constant**

The intensity of solar radiation outside the earth's atmosphere at the mean distance between the earth and the sun; equals

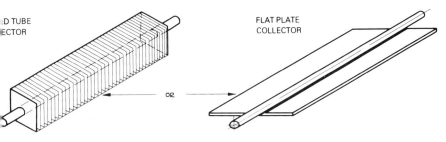

:D TUBE
ECTOR

FLAT PLATE
COLLECTOR

OR

2.9    Solar collector – finned tube or flat plate

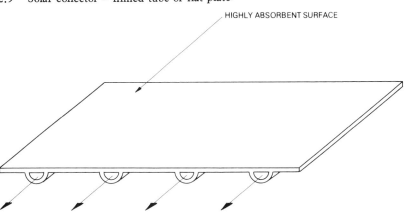

HIGHLY ABSORBENT SURFACE

2.10    Conventional flat plate collector

|                        | 1.353 kW/m² or 1.353×10³ J/sec/m² or 1.940/min/cm² or 428 Btu/hr/ft². |
|------------------------|------------------------------------------------------------------------|
| **Solarimeter**        | Instrument for measuring the solar irradiance, including both direct and diffuse components. Mounted horizontally, it will measure global irradiance. |
| **Solstice**           | The time at which the sun reaches its greatest declination, north or south. |
| **Spectral distribution** | An energy curve or graph which shows the variation of spectral irradiance with wavelengths. |
| **Stop cock**          | A 'shut-off' valve used for water circuit isolation and maintenance purposes. |
| **Stratification (thermal)** | Process used in heat store systems whereby the less dense hot water floats on top of the more dense cold water. |
| **Thermo-mechanical**  | The process of thermal expansion or con- |

|  |  |
|---|---|
| | traction of a temperature-sensitive metal or fluid employed to actuate the mechanism of a control device. |
| **Thermostat** | A sensor, electrical or non-electrical, measuring temperature fluctuations and activating a related control function. |
| **Thermosyphon** | A circulating system where a circulating pump is not used. Hot water, being less dense than cold water, rises to the top of the system and is replaced by cold water, in the process setting up a natural circulation of water through the system. |
| **Transparent reflective surface** | A surface coating which allows short wave radiation to pass through it while infra red (thermal) radiation is reflected. |
| **Ultra-violet** | A band of electromagnetic wavelengths adjacent to the visible violet $(0.10\,nm + 0.38\,nm)$, of particular interest in solar energy application since various unprotected materials change in colour due to reception of ultra-violet radiation; if the materials are unstable, they are then likely to suffer failure due to decay or fracture. |
| **User (utilisation) factor** | A factor employed in calculating the actual benefit derived from a solar hot water supply system based upon actual user habits (See chapter 6). |
| **Water (hardness)** | A term used to measure the extent of water supply scaling impurity commonly caused by calcium and magnesium salts. Derived carbonates and bi-carbonates are generally responsible for 'temporary' hardness, whilst chlorides, sulphates and nitrates contribute to 'Permanent' hardness of water. |
| **Water (scale)** | Solid encrustation from water impurities sometimes found inside water pipes and on heat exchanger surfaces – called 'furring' – causing water flow restriction and blockage. |
| **Water treatment** | A process of rendering scaling and corrosive impurities in water harmless to the |

vessels, pipes and components of a system carrying water or water mixtures.

**Weathering**  Means of preventing water or moisture penetration into building. Lead is a common roofing material sometimes used as 'flashing' or weathering strip between a collector and rooftiles. Other materials used are synthetic compounds, aluminium and rubber-based strips.

# 3

# SOLAR COLLECTORS

Cats sit on warm tin roofs because such roofs are comfortable, being good conductors of heat. A matt black metal roof with thermal insulation applied to the underside is still warmer and constitutes a basic flat plate solar heat collector, Fig. 3.1.

The ability of a material to absorb energy from the sun and re-radiate only a small proportion of this heat is dependent on the nature and texture of the material. Good conductors such as copper, aluminium and steel are excellent absorbers when suitably treated. Plastics are not such good conductors and they are therefore used only, almost exclusively, for swimming pool applications where resistance to chemical attack is required and the working temperature is low.

3.1   Cat on hot tin roof

The function of a solar collector designed for the conversion of energy from the sun into heat depends on the selected shape and method and on the arrangements for the prevention of heat losses from the collector surfaces.

A wide range of different configurations and types of solar collectors is available. It is therefore essential for the intending user to understand the relevant advantages and drawbacks of the different collectors before making the choice for a particular application.

Solar collectors divide essentially into two basic categories: flat plate collectors (most widely used at present) and others.

### FLAT PLATE COLLECTORS (LIQUID TYPE)

Fig. 3.2 shows a section taken through a typical flat plate collector. It will be seen that such a collector comprises:

Absorber plate and fluid lines
Thermal insulation under the plate
Cavity above the plate
Transparent cover, usually of single or double glazing
Enclosing and supporting frame
Optional: selective coating.

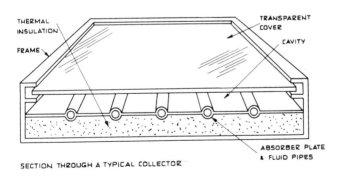

SECTION THROUGH A TYPICAL COLLECTOR

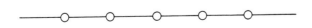

TUBE - IN - STRIP

3.2   Section through typical wet type flat plate collector

A well-designed and constructed collector will have only minimal heat losses, so that up to 85% of the solar radiation entrapped
by the glazing is available for heating of the water (fluid) circulating through the collector pipes. The hot water is circulated by
thermosyphon or pump assistance between the collector and the
heat store.

Transparent covers may be of glass or plastic, arranged in
single, double or multiple layers to provide a greenhouse effect.

Glass has a higher transmittance factor than most other transparent materials and is thus favoured. Glass does not suffer from
corrosion or ultra-violet instability, it has a considerable life, is
reasonably cheap and, with care, easily handled.

The specification of a typical good quality transparent glass
cover is:

> *Materials:* 3 to 4 mm glass plate. (The thicker glass will stand
> up to rougher handling; 3 mm thickness is widely
> used.)
> *Solar radiation transmission:* 92%
> *Solar radiation absorption:* 2%

There should be no degradation of performance with the passage of time, although periodic hosing-down or washing of the
cover is required for prolonged optimum performance. 3 mm
thickness of glazing is considered the minimum; using thinner
glass will reduce stability (requiring more frequent supports) and
increase breakages in transport and site handling.

Various alternative plastic covers have also been developed,
such as translucent reinforced polyester/glass fibre sheeting with
external weather surface coating. This type of cover is particularly
useful for certain applications, being impact-resistant, lighter than
glass and less likely to fracture due to uncompensated thermal
movement.

The thermal insulation backing must be adequate in thickness
and securely applied. Except in special cases, a minimum thickness of 60 mm (2 in) of closely packed glassfibre, fire retardant
polyurethene or other suitable insulant is necessary. Insulating
materials which are affected by moisture, chemical or biological
attack must not be used.

Certain special paints cannot be successfully used as absorber
coating with a double or multiple cover assembly, as they cannot
withstand the high temperatures involved and tend to 'smoke' the
cover. Under such conditions, a selective coating should be

applied. The situation can be reviewed when there has been a major advance in paint technology for solar applications. A good epoxy paint may be used under a single transparent cover.

Selective absorber coatings, such as black-chrome, are relatively expensive to process and therefore they are not commonly adopted, except with collectors intended for heating and air conditioning application. Lesser selective coatings are now more commonly used in the manufacture of hot water collectors.

More sophisticated versions of the basic flat plate collector include:

The use of more than just one single transparent cover
The incorporation of multi-layer interference coating copper oxides, meshes with aluminium film etc, to provide 'selective' absorber surfaces
The extension of the fins from the fluid tubes, cavity honeycombs, partial vacuum and other devices to reduce cavity heat loss convection and to improve absorption.

**Methods of absorber fabrication**
Figs. 3.2, 3.3 and 3.4 illustrate the method commonly employed in the manufacture of the absorber collector components. These include the tube-in-strip method; two corrugated pressed mild steel sheets; the roll-bond process; closely spaced black tubes; flat and ribbed sheet, ribbed sheet and close-fitting tubes; tubes welded into headers; rigid plastic tray and flat metal top; finned tubes; vertical fins on flat plate; V-shaped corrugated plate.

Copper has achieved a dominant role in the manufacture of solar collectors for the following important reasons:

Good conductance properties
Natural oxidisation tends to improve absorptance quality (untreated surface)
Has the quality of high resistance to the more common forms of chemical attack
Hot water supply represents the largest solar market and copper already dominates general hot water plumbing needs – the compatibility of materials is most important
Copper is fairly easily manipulated and repaired
Copper pipes have a smooth bore, making the adherence of scale more difficult
Copper is a relatively cheap material.

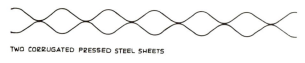

TWO CORRUGATED PRESSED STEEL SHEETS

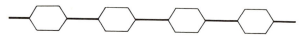

ROLL BOND PROCESS

CLOSELY SPACED BLACK TUBES

RIBBED SHEET & CLOSE FITTING TUBES

TUBES WELDED INTO HEADERS

## 3.3 Absorber collector types

Copper suffers from some *disadvantages* which must be recognised and compensated for. These are mainly its ductility (lack of stiffness) and susceptibility to external damage.

Possibly the most mass-produced heat-exchanger today is the steel panel radiator which is formed by expensive machinery employing the techniques indicated in Fig. 3.4. Although copper may be suited to these production methods, they have not been widely adopted for that material.

Closely spaced tubes, as shown in Fig. 3.3, have a limited use. Some manufacturers simply coil soft copper tubing into a flat-plate collecting box – this offers the obvious disadvantages of high cost and high hydraulic resistance.

The technique of mechanical bonding is sometimes used for the application of copper fins to copper tubes, but because of the lack of 'spring' in copper, it must be applied with care; a simple sheet

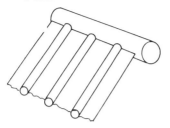

TUBES WELDED INTO HEADERS

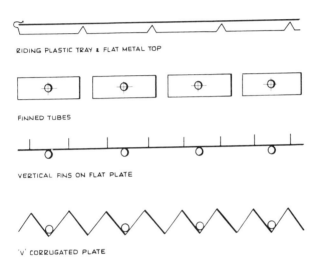

RIDING PLASTIC TRAY & FLAT METAL TOP

FINNED TUBES

VERTICAL FINS ON FLAT PLATE

'V' CORRUGATED PLATE

3.4   Absorber collector types (continued)

metal folding machine may be used for this purpose. This technique and that shown in Fig. 3.3 are widely used with hardened aluminium fins, but suitable protection must be applied at the interfaces between the dissimilar metals (copper and aluminium). The application of aluminium fins to copper tubing is best carried out by purpose-made and expensive machinery; it is probably cheaper to buy ready-made finned tubes than to attempt this operation in a small workshop, with limited facilities.

Perhaps the most popular fabrication technique is that shown in Fig. 3.2: the tube-in-strip. Sections of single tube-in-strip are commonly supplied in an overlapping fashion to enable small operators to make up solar collectors to their own designs.

The method of jointing tubes to headers, finned or otherwise, as shown in Fig. 3.4, is widely favoured and may well be carried out in a small workshop without much expenditure on tooling. It is best to set out the tubes on a frame (which could be of timber) to ensure the correct alignment and grading of the headers. A simple swageing tool may be used to provide an overlapping joint, finished with high grade silver-solder; butt-jointing is prone to damage in transit and to leakage.

The configuration of the absorber plates in Fig. 3.2 is designed to occupy most of the cavity between the transparent cover and the absorber to reduce cavity convection heat losses.

Plate temperatures between tubes will be higher than at tubes producing a temperature gradient. Such peak temperatures increase heat losses and may be reduced by choosing:

Material of better conductivity
Greater plate thickness, giving better conductance
Closer spacing of tubes
Turbulent, rather than laminar fluid flow
Good bonding of tube to plate to improve the heat transfer coefficient from plate to fluid.

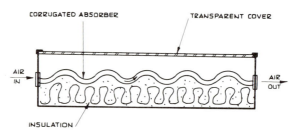

CORRUGATED ABSORBER                    TRANSPARENT COVER

AIR IN                                          AIR OUT

INSULATION

A SIMPLE CORRUGATED SHEET AIR TYPE FLAT-PLATE COLLECTOR

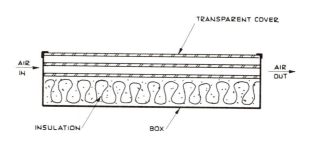

TRANSPARENT COVER

AIR IN                                          AIR OUT

INSULATION          BOX

TWO SHEETS OF GLASS OR RESISTANT PLASTIC AS AIRWAY

3.5   Dry type flat plate collectors

## FLAT PLATE COLLECTORS (DRY TYPE)

Air is much less suitable than water as a heat transfer fluid, possessing relatively low density and poor specific heat properties.

Nevertheless, in the USA most heating systems are 'dry', employing warm air *direct* heat transfer methods.

In use, the air to be heated is circulated between the collector and the heat store by a ducted fan system, incorporating dampers and controls. There are certain inherent advantages in this method, as air will not freeze, system leaks cause no damage and corrosion is seldom a major problem. Some collector forms are shown below.

Collectors using the configuration of Fig. 3.5 possess an operating efficiency of around 45% with a temperature difference of 65 °C (117°F) above ambient. Using a double pass system the air, passing through the outer duct first and then back through the inner duct, produces an efficiency increase of some 17% above that of a conventional double-glazed air heater.

Fig. 3.5 shows a simple corrugated sheet air type flat plate collector and a collector using two sheets of glass or resistant plastic as airway: Fig. 3.6 shows a typical air type flat plate collector with pebble bed heat store and Fig. 3.7 a contra-flow solar air heater.

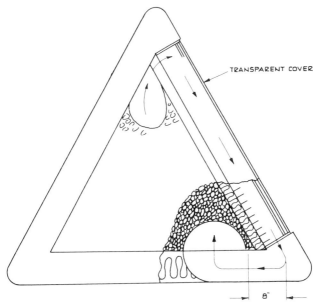

3.6    Air type collector with pebble bed heat store

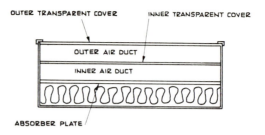

3.7    Air type flat plate collector showing contra-flow

STRUCTURALLY INTEGRATED COLLECTORS

There are a number of ways of incorporating flat plate collectors into the structure of a building. One such method, shown in Fig. 3.8, offers the advantage of visual integration and good weather-proofing. Such arrangements should be designed into the overall concept of the building or structure, rather than as an after-thought.

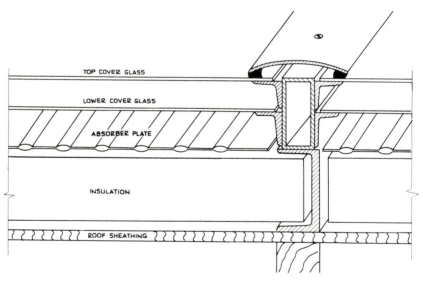

3.8    Structurally integrated collector

**Inter-truss collector arrangement**
A trend in modern housing is to provide flat plate collectors mounted between trusses spaced about 6 m (19.7 ft) apart, thereby enabling the builder to place patent glazing over the

assembly, improving appearance, facilitating maintenance and reducing the installation costs of the collectors.

This method of incorporating solar collectors under glazing appears at first sight to be an obvious method of applying such equipment to new housing; it avoids the precarious on-tile or expensive tile-removal type of collectors and eliminates the requirement for separate collector glazing.

Roof joists, rafters, and trusses are not designed to seal on to patent glazing. In consequence, any collector fitted between roof members without its own cover will suffer excessive heat loss from the absorber surface, materially and unacceptably reducing the collector efficiency. Until building construction methods are adapted for sealing solar collection installation, any collector mounted in this fashion must have its own cover. Fire protection of exposed timbers must also receive attention.

Solar radiation entering a collector (with own cover) which is mounted between roof members and under patent glazing must then be diminished by irregular spacing of double cover and certain re-radiation losses.

Between-joist application is obviously likely to be too expensive and may be discounted. The trend in modern housing, in any case, is to build with rafters spaced at 600 mm centres (19.7 ft) or with trusses at similar spacing centres. Inter-truss collector mounting must involve trimming members and some detailed attention to methods of collector support. Intermediate rafters provide a supported distance approximately equivalent to the conventional collector size and thus provide the most practical application method.

The siting of collectors under patent glazing certainly offers many desirable features, not least a more pleasing appearance; to achieve this end economically, the many inherent problems will eventually be overcome.

## FLAT PLATE COLLECTORS – A SUMMARY OF SALIENT FEATURES

The transmission co-efficient of the glazing or other cover must be as good as possible
Heat losses from the collector must be minimised
The absorber or flat plate coating must be of good quality and, if possible, selective, so that:

The absorption co-efficient is a maximum
The emission co-efficient for long-wave is a minimum

The operating temperatures must be kept as even as possible and at a usable level to avoid peaks and unnecessary consequent heat loss The materials employed in the fabrication of flat plate collectors for hot water supply application must:

be able to withstand exposed temperature differences of not less than 175°C (315°F)

be ultra-violet stable, suffering little or no decay from continual exposure to sunlight

possess fire resistance qualities complying fully with local authority building regulations

be able to support fixings

be able to withstand moisture, chemical and biological attack

be able to hold firm against wind and snow loadings

be fully accessible for maintenance purposes

be as light in weight as possible for their application

carry a manufacturer's guarantee against Fair Wear and Tear for a minimum period of 5 years.

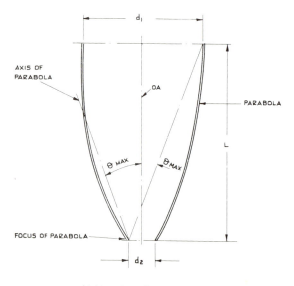

$$(d_1 / d_2) = (1/SIN\ \theta\ MAX)$$
$$L = (1/2)(d_1 + d_2)\ COT\ \theta\ MAX$$

3.9    Compound parabola

## FOCUSING COLLECTORS

Focusing collectors are not generally marketed commercially as low cost collectors for the more common, run-of-the-mill solar project. They are used extensively for heating and air conditioning applications because of their ability to operate at high differences of temperatures and to generate high grade energy.

Fig. 3.9, a compound parabola, illustrates the basic principles of the focusing collector and shows how light/direct beam radiation may be focused onto an object.

Fig. 3.10 shows the manner in which focusing technique may be practically applied to develop a semi-focusing collector from a flat plate collector.

In developing solar collection in this manner, it must be clearly understood that the focusing device performs well when there is light/direct beam radiation (clear sky). However, this device, having lost some of its absorbing surface, under-performs a flat plate collector in diffuse radiation conditions (bright cloudy sky). Figs. 3.11 to 3.15 show particulars of focusing collectors.

A further development of the focusing principle is to maintain as much absorber surface as possible to enable the collector to

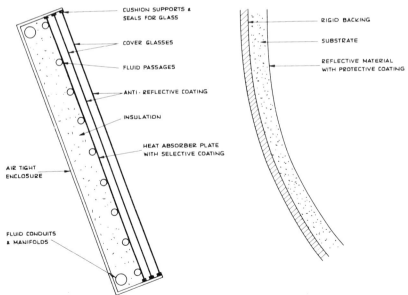

CUSHION SUPPORTS & SEALS FOR GLASS

COVER GLASSES

FLUID PASSAGES

ANTI-REFLECTIVE COATING

INSULATION

HEAT ABSORBER PLATE WITH SELECTIVE COATING

AIR TIGHT ENCLOSURE

FLUID CONDUITS & MANIFOLDS

RIGID BACKING

SUBSTRATE

REFLECTIVE MATERIAL WITH PROTECTIVE COATING

CROSS SECTION OF A TYPICAL FLAT PLATE COLLECTOR & FIXED CONCENTRATING REFLECTOR

3.10    Semi-focusing collector developed from flat plate collector

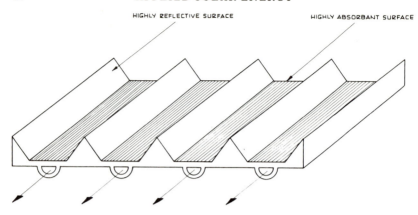

3.11    Moderately concentrating flat plate collector

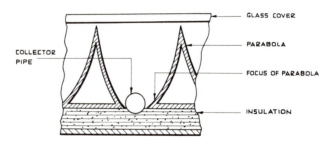

CONCENTRATING PARABOLIC COLLECTOR

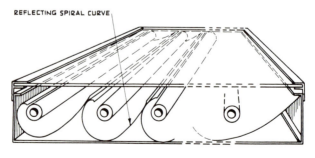

MODERATELY CONCENTRATING TRAPEZOIDAL COLLECTOR

3.12    Concentrating collectors

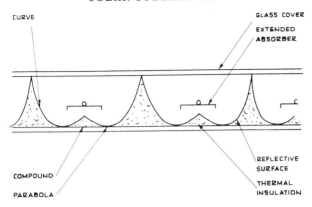

CURVE

GLASS COVER

EXTENDED
ABSORBER

REFLECTIVE
SURFACE

THERMAL
INSULATION

COMPOUND

PARABOLA

3.13    Concentrating collector assembly

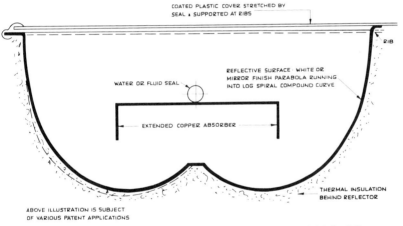

COATED PLASTIC COVER STRETCHED BY
SEAL & SUPPORTED AT RIBS

RIB

REFLECTIVE SURFACE · WHITE OR
MIRROR FINISH PARABOLA RUNNING
INTO LOG SPIRAL COMPOUND CURVE

WATER OR FLUID SEAL

EXTENDED COPPER ABSORBER

THERMAL INSULATION
BEHIND REFLECTOR

ABOVE ILLUSTRATION IS SUBJECT
OF VARIOUS PATENT APPLICATIONS

NOTE: BECAUSE OF CERTAIN SCATTERING OF IRRADIANCE THROUGH PLASTIC APPROXIMATED CURVE & WHITE
REFLECTOR IS SUFFICIENT. IF MIRROR FINISH REFLECTOR IS USED THEN CURVE MUST BE MATHEMATICALLY
CORRECT & GLASS IS A BETTER COVER THAN PLASTIC.

3.14    Parabolic focusing collector general principles

behave reasonably in diffuse radiation conditions, but at the same
time having the considerable advantage of focusing when there is
direct beam radiation.

One such device is shown in the photograph, Fig. 3.19. The
relevant particulars of this installation are:

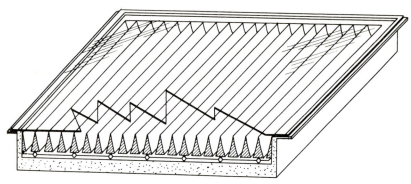

3.15    Concentrator in test panel

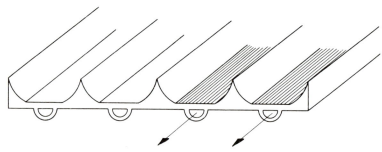

3.16    Better shape – compound parabola

| | |
|---|---|
| Collector array | 9.38 m² (101 ft²) |
| Heat store | 910 litres (200 gallons) |
| Location | London, England; latitude 51° 30′N |
| Maximum irradiance | 900 w/m² (285 Btu/hr/ft²) |
| System | direct heat transfer by thermosyphon |
| Savings | approximately 25% of annual auxiliary fuel cost. |

## COLLECTORS FOR SWIMMING POOLS

Flat plate collectors used for swimming pools do not, as a rule, incorporate a transparent cover.

The concept for solar collection in this application is entirely different to that for the conversion of solar energy into the heating of domestic hot water, for the following reasons:

a) Maximum pool temperatures are required at no more than around 30°C (86°F)
b) Pool water circulates through collectors utilising existing high-displacement water treatment pumps
c) Untreated metal water-ways cannot resist the chemical attack of chlorinated water.

In consequence, the 'art' has developed of using uncovered, all-plastic flat plate collectors, designed to provide only a nominal temperature rise, complete with the acceptance of a lengthy warming up period for the pool.

This simple technology has been found to be adequate for user needs, although many departures and unique methods of achieving the same results are now also being used.

The materials of a swimming pool collector must be completely compatible with the existing pool equipment and plumbing. The hydraulic resistance through the collectors should not create a significant overload on the pool pump.

Closed system collectors may be fully integrated to operate efficiently with a conventional auxiliary pool water heater, using no water return-to-pool pipes. It is also important that this type of collector, which may be located in any reasonable position above

3.17    Collector installation serving a toy factory

or below the pool, can fully withstand the pressure generated by
the pump in the swiming pool system.

Other types of collectors used for swimming pool applications
include:

a) Cascade no-pressure type – where the pool water is
released from the conveying pipe and allowed to fall by
gravity over, or through, the absorbers and back into the
pool

b) Under-paving-slab type – where the water pipes are
buried in, and under, paving slabs adjacent to the pool,
which transfer the energy and act as absorbers.

There are also various other kinds of swimming pool collectors,
but mostly these are based on the simple flat plate technology.

### Efficiency of flat plate collectors – to glaze or not to glaze

The graph in Fig. 3.18 typically shows the relationship for both
single-glazed and double-glazed panels. The graph information
clearly indicates that double glazing is only of benefit in circumstances where the mean water temperature is high and the collector
efficiency is generally low. One concludes that, in practice, single
glazing should generally be adopted, though there are certain

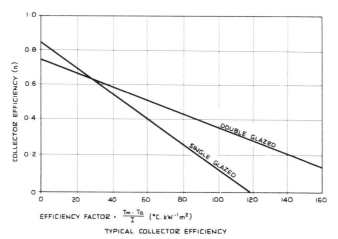

TYPICAL COLLECTOR EFFICIENCY

Tm : MEAN FLUID TEMPERATURE OF COLLECTOR °C
Ta : AMBIENT AIR TEMPERATURE
I  : INSULATION VALUE kW/m²

3.18   Collector efficiency – single- and double-glazed

specific applications which warrant the additional expense of providing double or even treble glazing.

Simple flat plate collectors commonly incorporate thermal insulation to reduce heat losses from the back of the absorber; additionally, the panel is glazed to reduce materially the convective and radiative heat losses. It will be readily understood that a collector operating close to ambient temperature will have only minimal convective heat losses; in such circumstances, the addition of thermal insulation to the back of the absorber can hardly be justified.

As well as reducing the convective and radiative heat losses, glazing reflects possibly 10% to 20% of the incident solar radiation – hence, there is a trade-off, or balance sheet, to be considered as between the gains and losses, respectively, which are attributable to the glazing. In the case of swimming pool collectors, an advantage would be gained from glazing only if the convective heat losses are known to be very high (such as would occur on very windy locations), but generally a bare, unglazed swimming pool collector will perform better than a glazed one.

At low angles of incidence, the transmission of glass or plastic materials becomes very low; an unglazed panel will receive significant amounts of insolation, even if its inclination and orientation is not near the optimum. It is much more important to ensure that the inclination and orientation of a glazed panel is correctly optimised.

Another possible disadvantage of glazed collectors arises at times when there is no flow of water through the collector (such as will occur when the heat store requirement has been satisfied or when the swimming pool is at the required temperature). In such a condition, the panel will reach very high temperatures when exposed to intense sunlight. In the case of a plastic panel, such exposure is likely to lead to rapid degradation of the plastic material. Where the collector is located on a roof, there will then also be considerable heating up of the building under the panel, unless extra thermal insulation is provided. The glazing may crack due to excessive expansion.

It is therefore concluded that glazing is of major advantage when operating at the higher water temperatures, such as would apply when heating domestic hot water. Glazing offers no advantage, and indeed certain disadvantages, when applied to collectors operating at low temperatures, as would apply to swimming pool systems. It has therefore been generally confirmed that standard glazed and insulated collectors are unsuitable for swimming pool applications.

## TESTING OF SOLAR COLLECTORS

It has been shown that the efficiency of any solar collector varies in accordance with a modified version of the Holtel-Whiller-Bliss equation (see chapter 2, Definitions).

It is very important that an internationally agreed standard testing procedure is eventually adopted. In the absence of such standards, and indeed some national standards, the layman finds it difficult to know with certainty whether one collector is better than another.

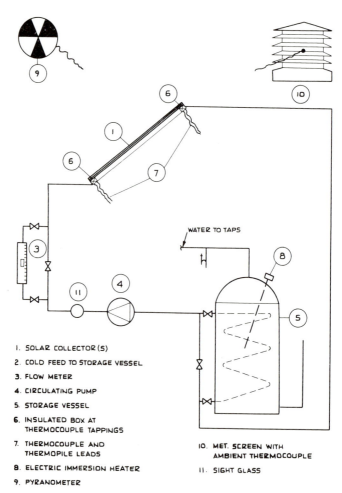

1. SOLAR COLLECTOR (S)
2. COLD FEED TO STORAGE VESSEL
3. FLOW METER
4. CIRCULATING PUMP
5. STORAGE VESSEL
6. INSULATED BOX AT THERMOCOUPLE TAPPINGS
7. THERMOCOUPLE AND THERMOPILE LEADS
8. ELECTRIC IMMERSION HEATER
9. PYRANOMETER
10. MET. SCREEN WITH AMBIENT THERMOCOUPLE
11. SIGHT GLASS

WATER TO TAPS

3.19    Schematic of collector test loop

Fig. 3.19 shows a typical arrangement of a test loop to establish the outdoor instantaneous efficiency curve in pseudo-steady-state conditions, broadly similar to the American Society for Heating and Refrigeration Engineers' standard. The effect of wind in this analysis must be considered together with transient conditions, (non-steady state) to obtain a realistic efficiency curve.

There are a number of indoor solar simulators, but results of comparable indoor/outdoor tests indicate, in the main, that indoor readings from experienced authorities are dependable.

In the absence of an established test procedure and test efficiency, it is always best to question the authority of the particular published efficiency curve related to the collector being considered for purchase.

# 4

# THE APPLICATION OF COLLECTORS

It is desirable to establish an orderly procedure for selecting a solar collector: to that end, the design of a solar heating system may be conveniently divided into the following two parts:

a)  The basic design and selection of the solar collector(s)
b)  The detailed engineering design of the features and the details of the chosen system.

This chapter is concerned with (a); chapters 5 and 6 deal with (b) and with the total concept.

### THE BASIC DESIGN

Primary consideration must be given to the economic viability of the proposal and to the likely pay-back period (see chapter 10). Energy conservation methods, other than solar energy schemes, should be reviewed to establish whether a *greater* economic advantage can be gained by alternative schemes, such as thermal insulation (see chapter 8).

One should not assume that an unlimited amount of solar energy can be collected at will. For any given location of a collector array there is only a limited amount of radiant energy available and this will suffer seasonal variation. In many areas, the collection of solar energy is worthwhile only during the summer, with some benefits during spring and autumn. Computer controlled studies by the Building Research Establishment of the UK Department of the Environment have shown that half the annual solar incidence in the UK occurs during the four months of May to August. It is therefore necessary to determine the availability of solar energy by reference to insolation levels, using tables and/or charts.

It is essential to establish how energy may be best utilised; whether the solar energy should deal with the whole or only with a portion of a particular heat requirement commonly met by a boiler or electric immersion heater. The extent by which solar energy can replace the conventional heat source may then be determined.

A suitable location must be available to accommodate the solar collector(s), of adequate area, correctly orientated, free from permanent shadow (such as that of an adjacent structure) and in close proximity to the solar energy system into which the collected heat is to be fed.

The optimum efficiency of heat transfer to the solar system is achieved by the correct selection of solar collectors as regards type, size and numbers of units to suit the particular application.

**Determination of insolation levels**
Weather stations' tables and charts list the 'maximum' and 'mean' values of insolation for differing locations and latitudes. They also advise in respect of clearness numbers, elevation above sea-level and any local environmental (micro-climatic) condition which may affect irradiance values. Weather stations are spread fairly widely apart and one must understand that the meteorological information can only be *strictly* accurate for areas in the close vicinity of the stations.

It is always desirable to consult also local experience, just in case the micro-climate in the locality of the proposed installation differs materially from that close to the particular weather station.

**Orientation**
In the northern hemisphere, for optimum results, it is best to locate the solar collector array facing due south or slightly west of due south. This is not always possible and a question then arises relative to the permissible limits of deviation from due south. Guidance is given below.

A roof or a building facing south-east, invariably loses the *afternoon sun* to shadow; the solar collection time is therefore reduced. A roof or a building facing south-west tends to lose the *morning* solar collection period. Thus, not much deviation from south may be tolerated with a *single* solar collector array.

Split collector arrays may be used to overcome orientation

difficulties: perhaps with one area facing south-east and another area facing south-west. Clearly, with such an arrangement more collector area is required and hence more cost is incurred, but an effect may be achieved thereby similar to that of a single collector array facing due south.

When a 'split' array is located to face towards east and towards west respectively, then loss of collection time is obviously greater. Common-sense must be applied to the siting of split collector arrays.

Figs. 4.1 and 4.2 illustrate inclined solar collecting surfaces, the object at any particular latitude being to optimise the variable seasonal angle of incidence. However, practical considerations must prevail, eg, if the optimum angle is 45° and a particular roof slope is 35°, then there is little point in materially increasing application costs merely to obtain a marginal advantage.

## Relationship of collector array to system

Any type of solar collector may *not* be freely applied to *any* system without prior thought; eg, it would be futile to install a collector with copper absorber for application to a swimming pool, where chlorinated water would quickly and inevitably corrode the metal surfaces.

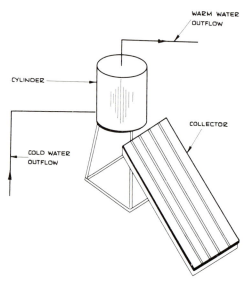

4.1   Single panel solar collector in assembly with roof-located hot water storage tank

**Table 4.1 Types of solar collectors**

| Type | Comment | Performance |
|---|---|---|
| **Fixed flat plate** | 1) Widely favoured for general and swimming pool uses<br>2) Low cost<br>3) Simple construction<br>4) Minimum maintenance requirement<br>5) Relatively easy to apply selective coatings | 1) Efficiency falls rapidly with increasing Delta (T)<br>2) Better than most other types in diffuse radiation<br>3) Output can be increased by providing extended surfaces (fins) |
| **Fixed semi-focusing** | 1) Increasing in popularity<br>2) Medium cost<br>3) *Highly polished* reflective surfaces require periodic cleaning<br>4) *White reflective* surfaces require less attention | 1) Efficiency not as good as flat-plate with low Delta (T)<br>2) Considerably less fall in efficiency than flat-plate with increasing Delta (T)<br>3) Performance in diffuse radiation less good than flat-plate conditions<br>4) Performance in direct radiation conditions better than flat-plate |
| **Fixed focusing** | 1) Generally used for specialised application<br>2) Medium/high cost<br>3) Periodic cleaning of surfaces required | 1) Poor *low* Delta (T) efficiency<br>2) Good *high* Delta (T) efficiency<br>3) Poor output in diffuse radiation conditions<br>4) Good output in direct radiation conditions |
| **Vacuum tube** | 1) A sophisticated product used for specialised purposes<br>2) Expensive<br>3) Periodic cleaning of surfaces required | Usually better efficiency than most collector types, both with low and high Delta (T) |
| **Tracking collectors** | 1) Sometimes applied to differing types of solar collectors for specialised purposes<br>2) Expensive<br>3) Require frequent and skilled maintenance | Normalising of incidence angle considerably increases efficiency |

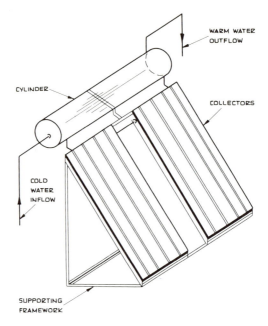

4.2     Twin-panel solar collector in assembly with horizontal hot water storage cylinder for roof location

The selection of collector type must be compatible with the intended system. Table 4.1 lists certain options in collector selection.

The overall concept of a solar collector array and system must allow close matching. Because of the difficulties and restrictions of collector accommodation, it is sometimes the collector array which selects the system. Basically, practical common-sense should prevail over purely theoretical notions.

**Fixings**
The adoption of correct fixing methods is crucial to the overall long-term success of a collector installation; the cost is also affected by the choice of fixings.

There are three major methods of applying solar collectors to a sloping tiled roof:

1)  With tiles removed
2)  With collector resting on tiles
3)  With absorbers under patent glazing.

The following well-tried procedure indicates how a collector with weathering flange and fairings can be integrated into part of a roof. Figs. 4.3 and 4.4 show a collector installation to a residence in South London, UK, to which this method was applied.

Step 1:  A tower scaffold and a long ladder were placed into position for use as a ramp to lift the collectors onto the roof.

Step 2:  An appropriate area of tiling was removed, this process being fairly simple. Suitably sized plastic sheets were kept to cover and protect the stripped area in the event of wet weather. The area of tile removal was larger than the collector – an extra strip of 0.3 m (1 ft) wide being stripped in each direction.

Step 3:  The most difficult step: to accurately position the collectors in relation to the existing tiling to avoid an unnecessary stripping of tiles. An added complication: the positioning of the flow and return pipes between heat store and collectors to ensure the absence of obstruction and full accessibility to the pipes after the roof has been reinstated.
The successful conclusion of Step 3 involved the appraisal of:
a) the relative positions of heat store within roof space
b) the relative positions of joists/rafters
c) the position of pipe runs in relation to angle and pitch of the roof to ensure correct pipe grading
d) difficult pipe joints adjacent to collectors with roof reinstated; insulation of pipes before tiling.

Step 4:  When the approximate position had been established, it was decided that one top corner of the collector array should be lined up 25 mm (1 in) from both a vertical and horizontal join in the tiles. It was found that the lines of the tiles were not straight or regular (this is not uncommon with older houses) and a considerable number of tiles had, in any case, to be cut since the 25 mm (1 in) gap at the top closed to virtually nothing at the bottom.

Step 5:  With the collector array positioned, the following two supplementary items were executed:
a) To avoid rainwater being driven under the tiles by strong winds, and over the edge of the collector

4.3   Focusing collector integrated with tiled roof

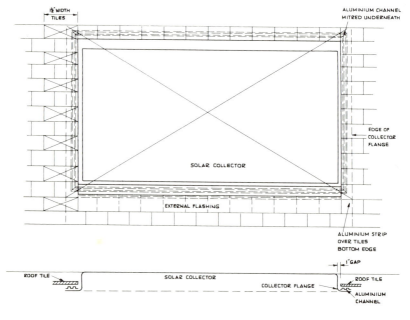

4.4   Focusing collector integrated with tiled roof (detail)

flange, aluminium flashing channel strip was
used.

b) A number of extra long tiles (1½ times the normal length) were fitted to finish alternative rows.
These wide tiles are clearly shown in Fig. 4.3.

It should be noted that the handling of solar collectors up a
make-shift tower and ladder is a hazardous affair, particularly
when executed during periods of strong winds. A handy and help-
ful neighbour can smooth this operation.

All reasonable safety precautions must be taken, including the
provision of hand-railing, extra long and safe ladders, etc, to
ensure the safety of everyone concerned in the operation.

Do-it-yourself enthusiasts should never undertake such installations single-handed.

## Loading

In setting up a roof-mounted collector application, one must consider the following:

a) roof loading
b) wind loading
c) snow loading
d) movement loading.

*Roof loading*

The weight of one square metre of solar collector may be 18 to
36 kg (40 to 80 lb); therefore, the total weight of the installation
on the roof may be in the order of 180 kg (400 lb).

The addition of the heat transfer fluid and pipe connections is
likely to increase the loading to about 227 to 272 kg (500 or
600 lb). Most well-constructed older houses will accept such extra
weight on their rafters but many more skimpily built modern houses
cannot carry such extra weight without modification. It is therefore always advisable to consult a competent builder or surveyor
before placing additional loads onto a roof.

*Wind loading*

In the case of an integrated roof collector (with tiles removed or
under patent glazing) wind loading is not usually a major concern.

However, with collectors placed on tiles, wind loading may well
be an important design factor. Local wind data should be
checked.

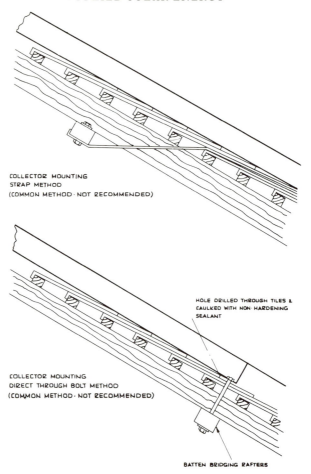

COLLECTOR MOUNTING
STRAP METHOD
(COMMON METHOD · NOT RECOMMENDED)

HOLE DRILLED THROUGH TILES &
CAULKED WITH NON · HARDENING
SEALANT

COLLECTOR MOUNTING
DIRECT THROUGH BOLT METHOD
(COMMON METHOD · NOT RECOMMENDED)

BATTEN BRIDGING RAFTERS

4.5   Collector mounting methods

Fig. 4.5 shows two commonly adopted methods of collector fixings by using straps and direct-through bolts, respectively.

The *strap method* offers a simple fixing, but may *not* be adequately strong and secure to hold down the collector under severe wind conditions. Should a signficant wind pressure difference occur across the collector, causing it to lift from the tiles, the wind exerts pressure on the underside of the box and there is then a real danger that the fixing straps, collector and tiles may break away from the roof.

The *bolt method* provides a much more rigid connection, but it allows neither for the unequal thermal movement of the collector and the roof, nor does it safeguard absolutely against excessive build-up of wind pressure across the collector.

Attention is drawn to the limitations of the 'strap' and 'bolt' methods and great care should be taken with the fixing details of such installation.

For new housing, on-tile collector application is unlikely and would seem to be unnecessary. It also is best to avoid on-tile fixing with existing housing, if at all possible. A meaningful attempt should be made to provide fairings or flashings so as to integrate the collector with the roof and thereby permit collector assembly and roof to move together without restriction, thereby lessening the effect of high wind pressure across and underneath the collector.

The vibration in the wind of large unsupported areas of collector covers, particularly when plastic is used, can cause difficulties; if the strength and rigidity of the cover is low, additional support should be provided.

The photograph 4.6 shows a typical roof on a house in Spain, on which it is virtually impossible to apply collectors 'on-tile'. The roof tiles are traditionally loose-fitting to cope with the large thermal movements caused by the wide climatic swings in air temperature. The integration and weathering of a collector in such circumstances is difficult and requires a measure of ingenuity.

4.6   Collector panels mounted on a tiled roof in Spain

The *suspended method* of collector fixing adopted in this case incorporated adequate anchoring for wind pressure effect without interfering with any roof movement. There are no weathering problems and a slight bonus is offered by the better inclined surface.

There is no valid excuse for using a 3 mm ($\frac{1}{8}$ in) glass cover over an unsupported area greater than one square metre at high northerly and southerly latitudes, as snow loading, even at some lower latitudes, will break such glass. The collector cover should essentially be tough, self-supporting and well able to withstand snow, ice and wind loads.

Wall fixings for collectors may generally be more robust than fragile roof assemblies. Great care must be exercised in the determination of the wind, weight and movement loadings of such supports and brackets.

Modern factory roofs have very little tolerance for additional weight and loading. Indeed, this aspect of commercial and industrial buildings emphasises the need for careful assessment and calculation.

Satisfactory swimming pool collector fixings are quite often most difficult to provide, usually because the unsightliness of the collector array suggests a location removed from made-up ground; this aspect must be considered when estimating or preparing such installation.

It is not advisable to hang swimming pool collectors of extra large area onto fixings which may be flimsy and ill thought-out. Block or concrete foundations or posts must be used. These collectors must also be stiffened against 'whip' in high winds. It is sometimes better to hinge this type of collector, so that 'fastening-down' may be effected in winter and at other times when the pool is not in use.

The success of any collector application will depend largely on the adoption of a correct installation procedure. However sophisticated a collector, the installer will find himself in difficulties if weathering between collector and roof is deficient, if positioning is poor and if weight, wind, snow and thermal movement loadings are too great.

### Summary – collector specification

Collector type and size must permit correct fixing in respect of wind and snow loadings, thermal movements and hostile environment.

Solar collectors should be chosen for their reliability, durability performance, need for minimum maintenance, low cost and particularly for their suitability for the application.

Solar collector arrays should be visually pleasing; at the worst, their appearance should not offend.

A thorough check should be made to ensure that the collector application will not infringe local authority regulations, legislation, insurance, etc, aspects (see chapter 10).

### Selection of collector – an example

Solar domestic hot water heating is contemplated for a three-person household located near to latitude 49°N. The house has a south-facing roof slope and adequate space for accommodation of all the required equipment and facility for integration with the existing hot water service and gas fired boiler installation.

Calculation and chart reference provide the following information:

a) Total maximum value of direct and diffuse spectral irradiance for roof incline $35° = 780$ w/m$^2$ (220 Btu/hr/ft$^2$)
b) Discounting December/January and early morning/late afternoon high incidence angles, the average useful insolation value $= 400$ w/m$^2$ (127 Btu/hr/ft$^2$)
c) Daily domestic hot water requirement $= 180$ litres (39.4 gallons)
d) Normal daily amount of heat required from auxiliary (back-up) heater (allowing certain thermal inefficencies) to maintain heat store at $60°C$ ($140°F$) $= 13$ kW
e) User/utilisation factor is determined at 0.7
f) Average heat input required from solar system is 9.1 kW
g) If average solar collection efficiency is 50%, then the required area of collector surface is 4.8 m$^2$ (52.2 ft$^2$).

#### TYPICAL COLLECTOR REQUIREMENTS

The following information is intended as a guide to the determination of solar collector sizes for typical domestic situations:

3 to 4 person household: 3 to 5 m$^2$ (32 to 54 ft$^2$)
5 to 6 person household: 4 to 8 m$^2$ (43 to 86 ft$^2$)

It might also be shown in the foregoing example that the solar energy contribution represents only about one third of the conventional (auxiliary) energy bill for the period in question. For a two-person household and, particularly, if there is a poor user factor, the capital investment in a solar collection system is difficult to justify at present levels of fuel costs and installation expenditure. There are also, of course, other considerations relating to locations with greater periods of sunshine and better user factors.

Referring to the foregoing example, the matter of possible overheating (boiling of the water) may be raised. Thus, if a maximum of 780 w/m$^2$ (247 Btu/hr/ft$^2$) is being converted at a 50% efficiency by an array of 5 m$^2$ (54 ft$^2$) over a continuous period of a few hours with an already hot heat store and little or no water usage, then there is a real danger of overheating, so that overheat precautions must be taken (see chapter 6).

It is desirable to somewhat over-size a solar collector array, rather than undersize it, as the latter is more cost-sensitive.

Although generally during December and January, and to some extent in November and February, insolation values are low at the more northerly latitudes in urban areas, elevated locations with near vertical collector surfaces may even then enjoy some useful solar collection periods.

**Selection**
Table 4.1 is intended as a guide in the selection of collector type.

### SUNLIGHT AVAILABILITY PROTRACTOR

The data of sunshine availability given in the usual weather offices' tables for different localities are very detailed. A quicker method for the evaluation of the availability of sunshine employs a sunlight availability protractor.

Average hourly values of sunshine are superimposed on a sun-path diagram (see curves spanning between the inner concentric circles), representing the average hourly probabilities of available sunshine in the particular locale being considered, as obtained from the weather bureau. Alternatively, the values can be interpreted as the overall expected durations of sunshine for each hour. The actual number of times within the hour that the sun may shine cannot be obtained, but this information is not normally required. The protractors shown as examples in Figs. 4.7 and 4.8 are based on recorded data of weather conditions in the London (UK) area, but can be used also for other locations, provided that the latitude and the climatic patterns are not *significantly* different.

The figures around the outer circle represent azimuthal angles and the directions N, E, S and W are clearly indicated. The

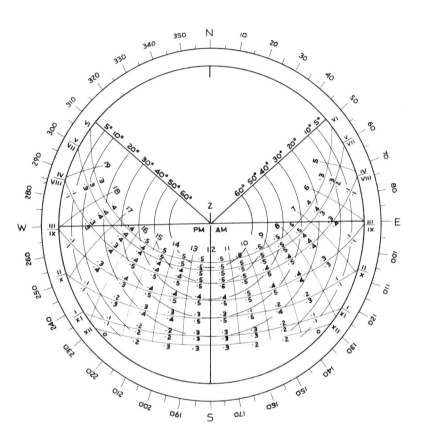

SUNLIGHT AVAILABILITY PROTRACTION
AVERAGE HOURLY PROBABILITY BASED ON
LONDON WEATHER CENTRE DATA SUNPATHS
FOR LATITUDE 51·5N ON 21st OF EACH MONTH

4.7   Sunshine availability protractor

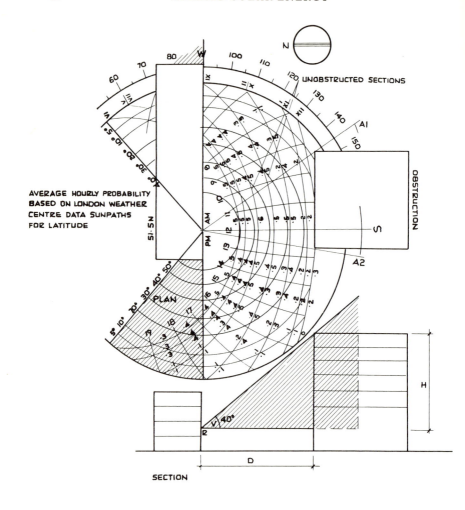

4.8    Use of sunshine availability protractor

curves which run from one side of the thick inner circle to the other represent projections of the sunpaths for the 21st day of each month. Each month is annotated in roman numerals between the thick circles. Most of these curves represent the sunpaths for two months and the sunshine durations are marked above, or below, the curve corresponding to the month.

The inner concentric circles represent solar altitude above the horizon, the centre of the diagram being the zenith. The sunpath projections are divided into solar hours and these are indicated above the top sunpath curve.

**Use of protractor**
It is worthwhile making a transparent version of the protractor, which can then be applied directly to plans and sections of buildings and building complexes. Relevant azimuthal angles (A1, A2) can be drawn on the plans and the sunlight availability protractor superimposed. The protractor must be placed so that its centre coincides with the point of interest or reference point (R), and so that its orientation coincides with the orientation of the building being investigated.

If the top of the obstruction subtends an angle (V) at the reference point of more than 62°, plans only are required to obtain the sunlight durations. However, where the angle subtended (V) is less than 62°, the months on which the obstruction will prevent the insolation of the reference point (R) can be found from the relevant solar altitude circles on the sunlight availability protractor. Probable durations of sunshine for the 21st day of each month can be obtained by summating the hourly values in all the unobstructed sectors.

For example, from Fig. 4.8 it can be seen that, say, on the 21st February the sunlight probability at the reference point (R) will be approximately 1.4 hrs, whereas the maximum possible duration of sunlight at that point is about 7 hours.

*Reference:*
R. G. Courtney, *A Computer Study of Solar Water Heating,* Building Research Establishment Current Paper, July 1977

# 5

# SYSTEMS:

# HOT WATER SERVICES

Possible arrangements of solar collection systems are numerous and varied; some of the more important ones are described in chapters 5 and 7 to illustrate the total concept.

The viability of a system depends upon two basic factors: supply and demand, or the availability of sunshine and the ability to use varying grades of collected energy effectively.

Generally, sunshine is more plentiful the closer the system is to the equator, although there are exceptions. In the more extreme northerly and southerly latitudes (in excess of 50°), the economic viability of solar energy systems is more difficult to prove; consequently, in these more exacting conditions, system technology and efficiency attains greater importance.

Even so, good engineering in system component assembly is essential, hand in hand with a basic understanding of the limitations and constraints imposed by local conditions, varying latitudes and utilisation factors.

## DOMESTIC HOT WATER SUPPLY

The daily bath is usually taken at a temperature of about 43.3°C (110°F); the system water is commonly circulated and stored at 60°C (140°F), this being an economical temperature level without incurring the risk of severe accidental scalding. School ablution water systems usually operate at a somewhat lower temperature of 55°C (131°F), or below.

Low-cost solar collector technology provides approximate

energy grades of 60°C (140°F) to 65°C (149°F). Solar-assisted domestic hot water supply has therefore become the first and foremost solar industry as the all year round, summer-and-winter demand for hot water enhances the viability of this solar energy application.

### DIRECT SYSTEM OF HOT WATER SUPPLY

The general arrangement of a typical direct hot water system, such as may require the addition of solar collectors, is shown in Fig. 5.1. In this, there is a direct primary connection between the hot water boiler and the hot water storage vessel; all water flowing to the outflow taps/fittings passes first through the boiler waterways.

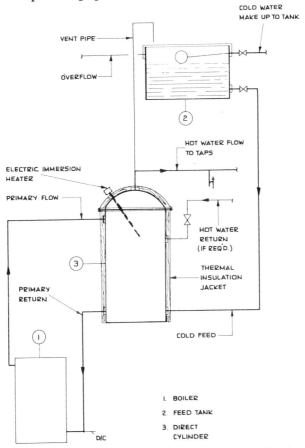

5.1    Direct system of hot water supply before applying solar collection

This arrangement is cheap to install but suffers from these major disadvantages:

a) Fresh make-up water flows into the installation in the same quantity in which hot water is drawn off and, with most waters, this make-up contributes either to the formation of scale in the boiler waterways or to their wastage by corrosion. Therefore, the hot water service boiler requires periodic inspection (with associated disruption to hot water supply) and cleaning of the internal waterways, as well as more frequent replacement.

b) The construction of the boilers must incorporate easy access for cleaning of the internal waterways, thereby imposing serious limitations on the design of the boiler and, in particular, on the amount of secondary heating surfaces which can be provided, especially in boilers of cast iron construction. The resultant shortfall of total boiler heating surface raises the exit flue gas temperatures and materially lowers the boiler heat exchange efficiency.

### INDIRECT SYSTEM OF HOT WATER SUPPLY

In the indirect system, the boiler water does not flow to the draw-off taps, there being a closed circuit between boiler and hot water storage heater battery, Fig. 5.2. The boilers may therefore be selected from a wide range of efficient closed-circuit heating boilers. Circuits for indirect systems are shown in Figs. 5.3 and 5.4.

Fig. 5.3 shows a convenient arrangement of imposing a solar collector upon an existing indirect system. The water in the intermediate storage tank 5 is heated by a heat exchange coil which forms part of a primary circuit; in locations where there is a risk of frost, the primary circuit will be filled with anti-freeze solution which will circulate between the collector and the heat exchange coil. Hot water from storage 5 is fed to the 'cold' end of the conventional hot water cylinder/tank 14. Since the primary circuit is a closed loop, it must be provided with an expansion tank or other feature for accommodating the volume changes as the water heats and cools. A control unit switches off the pump when the outlet temperature of the collector approaches that of the storage cylinder to prevent heat being lost from the storage to atmosphere.

The arrangement shown in Fig. 5.4 is suitable for a single residence or for small commercial premises. The roof-located collectors are piped up in parallel and form part of the primary circuit which, in colder climates, would be charged with anti-freeze solution. The fluid is pump-circulated between the collector panels and the heat exchange coil in the storage cylinder/tank. Control unit 7 with associated sensors switches the pump on when the outlet temperature of the collector exceeds the storage temperature by a pre-set differential.

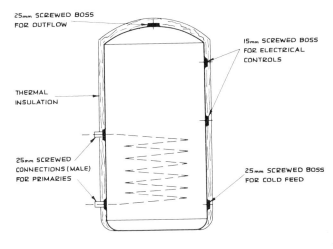

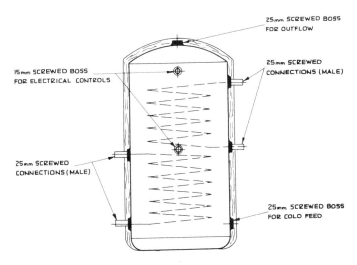

5.2   Typical hot water storage cylinders with heat exchange coils

When the collector outlet temperature has fallen and once more approaches that of the stored water, the pump is switched off. To allow for the expansion of the water on heating in the closed primary circuit, the circuit incorporates the expansion vessel 9 and associated safety valve 12, as well as appropriate topping-up facility.

Fig. 5.5 shows the previous scheme adapted to the requirements of a multi-apartment building. It will be noted that each apart-

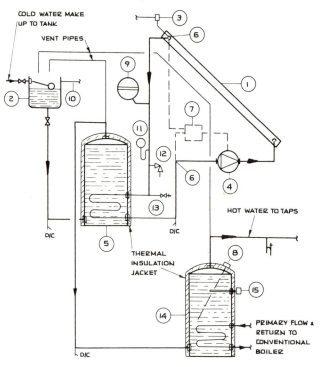

| 1. COLLECTOR(S) | 9. EXPANSION VESSEL |
| 2. COLD FEED TANK | 10. TANK OVERFLOW |
| 3. AUTO AIR VENT/VACUUM BREAKER | 11. PRESSURE GAUGE |
| 4. CIRCULATING PUMP | 12. SAFETY VALVE |
| 5. STORAGE VESSEL | 13. PRESSURE FILL |
| 6. TEMPERATURE SENSORS | 14. CONVENTIONAL INDIRECT CYLINDER |
| 7. ELECTRICAL CONTROL UNIT TO PUMP | 15. CONVENTIONAL THERMOSTAT |
| 8. ELECTRICAL IMMERSION HEATER | |

5.3   Schematic of an indirect system to which a solar collector system has been added

ment has its own separate hot water storage cylinder, the coil of each cylinder being connected to a common primary circuit. For standby purposes, each cylinder is shown fitted with an electric immersion heater. An alternative to this arrangement would be the provision of one common hot water storage to which all the apartments connect. A choice between the two types of systems depends on numerous factors, such as deliberate limitation of hot water useage per apartment (possible with separate cylinder system), arrangements for running secondary and primary pipes,

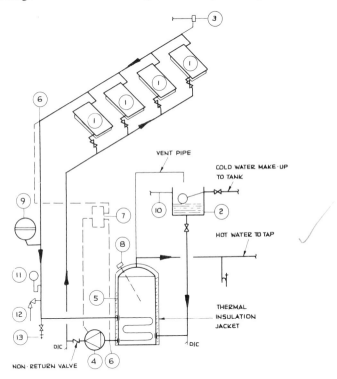

1. COLLECTOR(S)
2. COLD FEED TANK
3. AUTO AIR VENT/VACUUM BREAKER
4. CIRCULATING PUMP
5. STORAGE VESSEL
6. TEMPERATURE SENSORS
7. ELECTRICAL CONTROL UNIT TO PUMP

8. ELECTRIC IMMERSION HEATER
9. EXPANSION VESSEL
10. TANK OVERFLOW
11. PRESSURE GAUGE
12. SAFETY VALVE
13. PRESSURE FILL

5.4   Schematic of indirect hot water system using multiple collectors and one cylinder frequency of immersion heater use (this would favour the multi-cylinder scheme), space for individual cylinders, etc.

Great care must be taken in colder climates to provide efficient thermal insulation to all externally located pipes and fittings.

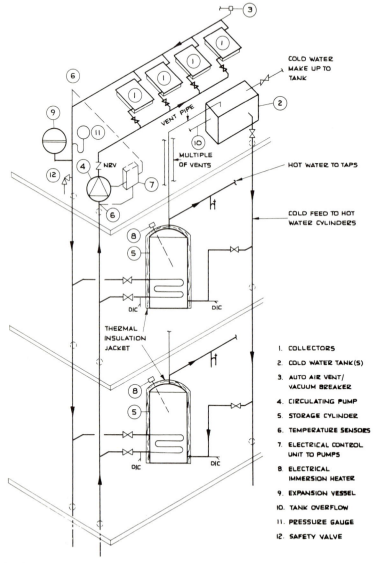

COLD WATER MAKE UP TO TANK

VENT PIPE

MULTIPLE OF VENTS

HOT WATER TO TAPS

COLD FEED TO HOT WATER CYLINDERS

NRV

THERMAL INSULATION JACKET

DIC

1. COLLECTORS
2. COLD WATER TANK(S)
3. AUTO AIR VENT/ VACUUM BREAKER
4. CIRCULATING PUMP
5. STORAGE CYLINDER
6. TEMPERATURE SENSORS
7. ELECTRICAL CONTROL UNIT TO PUMPS
8. ELECTRICAL IMMERSION HEATER
9. EXPANSION VESSEL
10. TANK OVERFLOW
11. PRESSURE GAUGE
12. SAFETY VALVE

5.5　Schematic of multi-collector, multi-cylinder indirect hot water system

The main advantage of an indirect hot water service system as compared with a direct system for solar collection lies in the major reduction of scaling and corrosion in the primary circuit. Also, the complete separation of the primary circuit from the draw-off water permits heat transfer media, other than water, to be used; although where anti-freeze solutions have to be used, the corrosive nature of these shortens the life of the system.

All types of solar panels installed in the northern latitudes are liable to suffer damage from frost; suitable precautions can be taken to prevent this by adding a suitable anti-freeze liquid to the circulating water in the closed circuit, ie, to the water that flows through the panel and the inner coil of the solar cylinder. Such a protective measure is suitable for all system layouts, with the exception of a solar water heating system with an open circuit.

The most widely used anti-freeze liquid for solar applications is Ethylene Glycol. To prevent internal corrosion of the pipe surfaces or channel surfaces in the case of 'open' types of solar panel, especially if aluminium is used in the construction, a corrosion inhibitor, such as sodium nitrite or benzoate, should be added and can be purchased, already mixed with the anti-freeze, as 'inhibited' ethylene glycol. A good example of this is 'Smith's Bluecol Universal', an anti-freeze liquid made especially for the protection of aluminium car engines to prevent freezing under normal winter conditions in the UK.

Ethylene Glycol is slightly toxic and is likely to contaminate the domestic water to the premises in the event of a leak in the heat exchange coil; its use should therefore be controlled carefully. The use of non-toxic propylene glycol is recommended in preference to ethylene glycol. The recommended proportions of glycol to water in an anti-freeze solution is listed in Table 5.1.

An anti-freeze solution in contact with air in an open panel or through the header tank in the case of a closed panel, is inclined to oxidise. Regular inspections should be made to check its condition.

Although much potential trouble is avoided by the choice of an indirect system, it is advisable to add a scale inhibitor to the primary circulating water.

Where the water can come into contact with air, as in the case of an 'open' panel, an algicide should be added to stop the growth of algae and moulds on the surface of the glass covers. Apart from being unsightly, algae deposits reduce heat absorption.

**Table 5.1 Proportion of glycol and water in anti-freeze solutions**

| Minimum ambient temp.°C | Ethylene glycol | | Propylene glycol | |
|---|---|---|---|---|
| | % glycol | % water | % glycol | % water |
| −5 | 20 | 80 | 22 | 78 |
| −10 | 25 | 75 | 27 | 73 |
| −15 | 30 | 70 | 33 | 67 |
| −20 | 35 | 65 | 38 | 62 |
| −25 | 40 | 60 | 44 | 56 |
| −30 | 45 | 55 | 49 | 51 |

DESIGN PARAMETERS

Arguably, solar hot water systems might be designed to provide a resultant primary fluid temperature of 50°C (122°F), rather than of 60°C (140°F), so that the initial installation cost of a system might be lower and its viability improved.

If a flat-plate solar collector of 1 m² (10.8 ft²) size is filled with water and inclined towards the sun with high ambient insolation value, then the water under static conditions might conceivably heat up to boiling point after a period of time. Circulated through a system which incorporates a reasonably sized heat store, the water is unlikely to boil, unless there happens to be a prolonged period without hot water usage. In northerly latitudes, the maximum likely circulation temperature is about 70°C (158°F), or so, and the heat store will then level out at about 60°C (140°F). Higher temperatures are achievable (though not necessarily desirable) at more southerly latitudes.

Since the cost of the collector array represents perhaps only one quarter to one third of the total cost of the overall system, and since the difference between a system designed for 50°C (122°F) and one designed for 60°C (140°F) might be one or two square metres of collecting surface in five or six square metres (perhaps 30%), the difference in the *overall* cost will be only in the order of 10%, or less. Hence, one might as well be generous with collector surface.

High grades of energy (ie, higher circulating temperatures) also reduce the required heat transfer surfaces; it is therefore beneficial to design all components for operation and use at the highest practicable temperature.

## SYSTEM RELATION TO A WATER HEAT STORE

It may be shown that at latitudes of about 50°, where it is more difficult to prove solar collection viability, the greater the demand for hot water, the closer the system comes to being economically viable.

Given a poor utilisation factor, it is always better to increase the heat store capacity; this larger storage also offers an added safeguard against overheating.

Where mains water pressure is low and the refilling of a 140 litre (31 gall) heat store might take half an hour, it is desirable not only to increase (possibly double) the usual heat store capacity, but also to increase the back-up cold water storage capacity; eg, under *such* circumstances, for a 3-person household, the particulars would be:

Collector array: 4 m² (43 ft²)
Heat store: 280 litre (60 gall)
Cold water storage tank: 280 litre (60 gall)

The more common 140 litre (31 gall) heat store is just about adequate for a normal household, provided there is good utilisation and mains water pressure.

## SCALING AND CORROSION IN DIRECT HOT WATER SUPPLY SYSTEMS

The harder the water supplied to the installation, the greater the extent of furring, as the calcium and magnesium salts are precipitated.

When water is boiled in a kettle, this furs up after prolonged use.

The calcium salts are responsible for the formation of hard scale; the magnesium salts break down to form acid and cause corrosion.

Thus, the main concern with direct heat transfer solar systems is scaling and corrosion; indeed, this was a major factor in the failure of many of the earlier solar systems installed at the beginning of this century.

In those early days, rough iron pipes were used, later replaced with low-carbon steel (0.15 to 2.28% carbon) and subsequently followed by the galvanised pipe (zinc coated) to resist corrosion. All these pipes had this in common: after the process of corrosion had started, it inevitably travelled through the pipe section in a lateral fashion; because of the relatively rough inside surfaces of the pipes and bends, precipitated salts were able to form 'nodes' and build up to eventually either completely block the pipes or else to severely restrict water flow and so render the system useless.

Nowadays, copper is widely used and the above difficulties are largely avoided. Photograph 5.6 shows a badly scaled (furred-up) pipe. The smooth inside surfaces of copper pipes do not permit precipitates to adhere; these are flushed through the system, ending down the drain.

It is recognised that much more research and development is desirable in this area to improve the life and effectiveness of the direct system of solar collector application, as the advantages of direct system heat transfer have been quantified by a number of researchers.

A more general acceptance of the direct system (without the risks of scale, corrosion and frost damage) could significantly reduce the costs of solar systems. Such systems are shown in Figs. 5.7 and 5.8.

5.6    Photograph of badly furred (scaled) water pipe

Fig. 5.7 shows the circuit for a direct system which is applicable to domestic use. Water is heated by circulation through the collector and the hot water storage. When the temperature at the outlet of the collector falls within a degree or two of that in the storage cylinder, as detected by the temperature sensors, the control unit switches off the circulating pump. The water in the collector and in the associated high level pipes then drains into the cold feed/dump tank which fills up to the upper level 'A'; the tank capacity between levels A–A absorbs this amount of water before the ball valve refills the tank. The dump tank also serves as the cold water storage tank, maintaining the water at the lower level when the

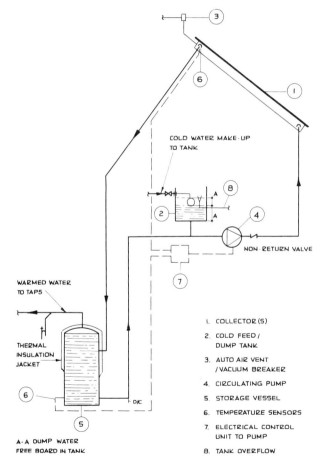

COLD WATER MAKE-UP
TO TANK

NON-RETURN VALVE

WARMED WATER
TO TAPS

THERMAL
INSULATION
JACKET

A-A DUMP WATER
FREE BOARD IN TANK

1. COLLECTOR (5)
2. COLD FEED /
   DUMP TANK
3. AUTO AIR VENT
   / VACUUM BREAKER
4. CIRCULATING PUMP
5. STORAGE VESSEL
6. TEMPERATURE SENSORS
7. ELECTRICAL CONTROL
   UNIT TO PUMP
8. TANK OVERFLOW

5.7    Schematic of direct hot water system with dump tank

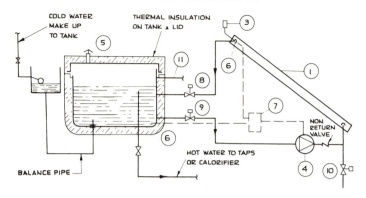

1. COLLECTOR (S)

2. STORAGE TANK

3. AUTO AIR VENT / VACUUM BREAKER

4. CIRCULATING PUMP

5. TANK VENT

6. TEMPERATURE SENSORS

7. ELECTRICAL CONTROL UNIT TO PUMPS

8.⎫ THERMOSTATIC CONTROL VALVE
9.⎭ SET TO CLOSE AT 1°C

10. THERMOSTATIC CONTROL VALVE
     SET TO OPEN AT 1°C

11. TANK OVERFLOW

5.8     Direct system of solar hot water supply – low silhouette

pump is in operation; this residual volume represents at least the volume of water required to fill the collector and associated pipes, should any of the dumped water be drawn off at the hot water tap(s). When the temperature difference rises to about 7°C (12.6°F), the control unit acts to switch on the circulating pump and the cycle is repeated. The position of the inlet to the storage tank is critical; it should be located one third from the top of the tank.

Fig. 5.8 indicates a similar, simplified system, which occupies relatively little vertical space and is therefore particularly suitable for locations on flat roofs or inside lofts. Similar to the circuit of Fig. 5.7, the pump is controlled to prevent cooling of the circulating water when solar conditions are unfavourable. However, an additional feature is provided for frost control (for use in locations where frost is likely) when the temperature of the collector and associated sensor falls below a pre-determined value – about 1°C (33.8°F); solenoid valves 8 and 9 close and solenoid valve 10 opens to drain the collector and associated pipework. When the temperature at the collector rises by about 7°C (12.6°F), the operation of the solenoid valves reverses and the system is made ready for solar heat collection.

The installer must check that the local Water Utility Company will not object to the dumping to waste of the collector water – the quantities are small, about 5 litres (1.1 gall) per typical single panel collector. One must also ensure that the ball valve of the cold water make up is not in contact with hot water – if there is a possibility of this being the case, then a separate make-up tank is required, as shown in Fig. 5.8.

Corrosion in the system is particularly undesirable, since an attack of localised severity may take place and remain undetected, causing much damage which may require repairs and perhaps expensive replacement of defective parts.

Corrosion in direct (open) systems may be due to a variety of impurities and salts in the water. The constitution of water is by no means universal; on the contrary, it varies greatly, depending on the water source. The installer is faced with a particular water and he should check whether this is especially aggressive or hard, so that appropriate precautions as regards materials and water treatment may be taken by him.

One form of corrosion can be, through ignorance, built into the system by the installer through the use of dissimilar metals having different electro-potentials, eg, copper and galvanised steel; electrolytic action will arise and the steel pipe will be attacked and corroded.

Aluminium pipes are only rarely used to convey water, because of the tendency of the aluminium to react in an accelerated fashion to dissimilar metals.

**Flux corrosion**

Copper pipes are quite commonly jointed with fusion jointing methods which employ solder and flux. The use of such flux and particularly the presence of excess flux may cause corrosion. It is important that all excess flux is removed immediately after jointing the pipes.

Low-fusion silver solder is known generally to have very limited chemical reaction with copper; hence the widespread use of fusion type water fittings. It is known however that in some areas there have been a number of failures. The tell-tale blue/green coatings on the inside of the tube and the eventual pitting and failure lead to only one solution following failure: the replacement of the fusion joints by compression type fittings.

In pipe systems manufactured in galvanised steel, it is important that the water temperature does not exceed 68.3°C (155°F) during the early life of the installation, before an eggshell thickness of scale has formed on the internal surfaces to protect the zinc coating. Failure to observe this precaution can result in the loss of the zinc coating and in consequent rapid corrosion. Apart from corrosion, the loss of the zinc coating in direct systems will result in the outflow of discoloured water at the taps.

Oil and grease must be prevented from entering the system, as, once admitted into the water, they tend to form into globules which adhere to the metal surfaces. Sometimes (at high temperatures) oil and grease char to carbon and insulate the heat transfer surfaces, leading to local overheating and subsequent bulging and collapse of the metal surfaces.

### Water treatment

Several methods of water treatment are commercially available; these may be modified or combined to suit the type of water and the operating conditions.

The symbol 'pH' (widely used in water analysis) is an abbreviation for potential of hydrogen; the pH scale of reference is numbered from 1 to 14.

Water with a pH of 7 is *neutral*. Water with a pH value of less than 7 is *acidic*; the lower the value, the more severe the acidity. Water with a pH value of between 7 and 14 is *alkaline*.

In cases where the installer is dealing with 'difficult' or hard waters, he should consult with a water treatment specialist as to suitable treatment, which may involve the installation of a water softener or of dosing equipment in the smaller installation.

The use of a base exchange water softener in an installation which includes galvanised mild steel pipes requires special care, even though it has been established that the local water is suitable for long-life usage with galvanised pipes (galvanised pipes should never be used with a water having a pH value of less than 7.2 or with a hardness of less than 210 ppm as $CaCO_3$ at 7.3 – at higher values of pH, a lower hardness is acceptable, such as 70 ppm at 7.9–8.5 pH). It is normally considered that the demarcation line between 'hard' and 'soft' waters is at 140 ppm as $CaCO_3$ (one ppm, part per million, is equivalent to 0.10 parts per 100,000; 0.058 grains per US gallon; 0.07 grains per imperial gallon).

The authors have had to deal with numerous cases of replacement of galvanised pipe installations necessitated by water being reduced in hardness below the above minimum hardness, at which condition the protective zinc coating is stripped from the internal metal surfaces and the bare steel is left vulnerable to inevitable corrosion attack by the softened water.

## THERMOSYPHON

Wherever feasible, in the smaller installations, it is desirable to circulate the water or primary fluid between the solar collectors and the heat store by gravity or thermosyphon action, rather than by use of an electric pump, as pumping adds to the cost, energy consumption and complexity of the system. Extra care must be taken in the design of a gravity system to ensure good circulation, particularly at low temperature differences.

The forces which induce the circulation by overcoming the resistance of the system components are due to the difference in density of the hot water in the flow pipe and the cooler water in the return pipe respectively. They are very slight – of the order of 2.5 N/m² (0.01 in wg). It is therefore essential to minimise frictional losses by careful design and attention to detail.

All thermosyphonic systems are self-regulating; the greater the energy received, the more vigorous the circulation, so that the collector operates at a steady efficiency. Reverse water flow in cold weather conditions is unlikely.

Not infrequently, gravity systems are installed in the hope that they will circulate. However, there is no valid reason why the installation engineers should not guarantee the correct operation of the system by careful design and by particular attention to the following guide lines:

1. Gravity systems do not, and *cannot*, exchange an optimum amount of heat, as by their very nature, circulation must be somewhat 'sluggish'.
2. Ample space must be available in the roof space for equipment.
3. The hot water cylinder (heat store) should be placed as high as is practicable above the collector and at a height of not less than 0.2 m (8 in) above the top edge of the collector.
4. Collectors should be located low on the roof slope to

provide maximum circulating head (H). If H is zero, then there can be no circulating pressure at any time, under any condition.

5. The flow pipe must be continuously graded upwards to correctly effect circulation and to continuously release entrapped air.

6. With any type of direct system, air will always be present in the circulating fluid, as water is drawn from the taps and continuously made-up by fresh aereated water.

7. The return pipe must be continuously graded downwards to correctly effect circulation and to facilitate drainage of the system at the lowest point.

8. Flow and return pipes should be kept as short as possible to minimise resistance to water flow. All fittings to be of the long-sweep and long-radius type.

9. If close proximity of collector and heat store cannot be achieved, then maximum possible 'H' must be established and frictional resistance reduced by increasing the pipe sizes within the primary circulation.

10. With open-top heat stores, when water is drawn off at the taps, there will be a fall in the water level, which may temporarily uncover the primary flow inlet connection into the heat store until the cold water ball valve has had time to make-up and refill the storage. Such a situation should be avoided, if at all possible, in the design of the heat store and its pipe connections.

11. Table 5.2 indicates recommended pipe sizes for the design of thermosyphonic systems.

A possibility exists whereby every precaution appears to have been taken to install an efficient thermosyphon system but, nevertheless, the water will not circulate effectively. One such likely case arises where a solar collector has been mounted in a horizontal position and because of tiling difficulties on the roof, the ladder/header is tilted in the wrong direction. The collector will then heat up and become hot, but there is no noticeable thermosyphon effect.

It is advisable, at the installation stage, to make certain that the collector has been tilted some 5° or more in the right direction and to test for adequate thermosyphon circulation, as soon as practicable and before the whole installation has been completed.

Another possible bar to circulation is the presence of unvented

**Table 5.2  Recommended pipe sizes for thermosyphonic systems**

| Area of solar collector | 3 m² (32.3 ft²) | | 4 m² (43 ft²) | | 5 m² (54 ft²) | | 6 m² (65 ft²) | | 7 m² (75 ft²) | | 8 m² (86 ft²) | |
|---|---|---|---|---|---|---|---|---|---|---|---|---|
| Total length of pipe use | 10 m (32.8 ft) | 20 m (65.6 ft) | 10 m (32.8 ft) | 20 m (65.6 ft) | 10 m (32.8 ft) | 20 m (65.6 ft) | 10 m (32.8 ft) | 20 m (65.6 ft) | 10 m (32.8 ft) | 20 m (65.6 ft) | 10 m (32.8 ft) | 20 m (65.6 ft) |
| Bottom of heat store above collector | | | | | Pipe size | | | | | | | |
| 0 | 25 mm (1 in) | 32 mm (1¼ in) | 25 mm (1 in) | 32 mm (1¼ in) | 32 mm (1¼ in) | 40 mm (1½ in) | 32 mm (1¼ in) | 40 mm (1½ in) | 40 mm (1½ in) | 50 mm (2 in) | 40 mm (1½ in) | 50 mm (2 in) |
| 0.15 m (6 in) | 25 mm (1 in) | 32 mm (1¼ in) | 25 mm (1 in) | 32 mm (1¼ in) | 32 mm (1¼ in) | 40 mm (1½ in) | 32 mm (1¼ in) | 40 mm (1½ in) | 40 mm (1½ in) | 50 mm (2 in) | 40 mm (1½ in) | 50 mm (2 in) |
| 0.30 m (1 ft) | 25 mm (1 in) | 32 mm (1¼ in) | 25 mm (1 in) | 32 mm (1¼ in) | 32 mm (1¼ in) | 40 mm (1½ in) | 32 mm (1¼ in) | 40 mm (1½ in) | 40 mm (1½ in) | 50 mm (2 in) | 40 mm (1½ in) | 50 mm (2 in) |
| 0.45 m (1 ft 6 in) | 25 mm (1 in) | 32 mm (1¼ in) | 25 mm (1 in) | 32 mm (1¼ in) | 32 mm (1¼ in) | 40 mm (1½ in) | 32 mm (1¼ in) | 40 mm (1½ in) | 40 mm (1½ in) | 50 mm (2 in) | 40 mm (1½ in) | 50 mm (2 in) |
| 0.6 m (2 ft) | 19 mm (¾ in) | 25 mm (1 in) | 25 mm (1 in) | 32 mm (1¼ in) | 32 mm (1¼ in) | 40 mm (1½ in) | 32 mm (1¼ in) | 40 mm (1½ in) | 40 mm (1½ in) | 50 mm (2 in) | 40 mm (1½ in) | 50 mm (2 in) |
| 0.75 m (2 ft 6 in) | 19 mm (¾ in) | 25 mm (1 in) | 19 mm (¾ in) | 25 mm (1 in) | 32 mm (1¼ in) | 40 mm (1½ in) | 32 mm (1¼ in) | 40 mm (1½ in) | 40 mm (1½ in) | 50 mm (2 in) | 40 mm (1½ in) | 50 mm (2 in) |
| 0.9 m (3 ft) | 19 mm (¾ in) | 25 mm (1 in) | 19 mm (¾ in) | 25 mm (1 in) | 25 mm (1 in) | 32 mm (1¼ in) | 32 mm (1¼ in) | 40 mm (1½ in) | 40 mm (1½ in) | 50 mm (2 in) | 40 mm (1½ in) | 50 mm (2 in) |

Note: The collector pipe connections are 15 mm (½ in) reduced from 19 mm (¾ in) at the collector high points in the collector-cum-pipe installations. All such high points, if indeed there must be such high points, have to be provided with accessible vent arrangements, incorporating needle valve and air bottle (the latter may be formed by a pipe fitting, such as a tee piece). In locations where the vent valve cannot be readily accessible, an extension pipe of small bore should be fitted between the top of the air bottle and the accessible valve bleed point.

The installer must check that all vent valves are tightly shut off after the venting operations. On pumped systems, the initial venting should take place with the pump switched off.

For particularly awkward venting situations, automatic vents are available and may be fitted – however, these are always a maintenance liability and require periodic checking to ascertain that they are not dribbling water.

### PUMPED SYSTEMS

The inclusion of a circulating pump within the pipe circuit between the solar collector and the heat store greatly increases the designer's options as regards location of the equipment, pipe sizes and control systems.

A typical pump for use with a solar system is shown in 5.9 and 5.10. The pump is of the canned rotor type, is very compact, has neither pump gland (vulnerable to leaks) nor driving belt. The pump should be fitted between isolating valves to permit its removal without having to drain the whole installation. Some pumps are manufactured with integral isolating valves. Some makes of pump are supplied complete with an adjustable bypass which permits the adjustment and regulation at site of the pump duty. In the event of a complete pump failure, it is usual to remove the pump and trade it in for a replacement unit.

Small pumps are usually specified by size of pipe connection, rather than by performance details. Each pump has its own characteristics as regards water flow and frictional head which the pump can overcome at the rate of flow, as indicated on the pump tables or graphs supplied by the pump makers. Cast iron construction is suitable for indirect systems, bronze construction for direct systems. In direct systems, a pipeline strainer should be fitted close to the pump suction to protect the small pump clearances from scale and dirt.

Using a pump, it is no longer necessary to have the heat store above the collector; it is practicable to locate the collector on the roof and the heat store at a lower level, even in a basement.

Control can be achieved by switching the pump on and off under detector sensing. The pump must be suitable for the available electric supply. Single phase electric supply selection is common for this application.

5.9   Glandless solar pump

The sizes of the pipes in the pumped circulation can be considerably reduced from those required for gravity circulation. It is thus easier to accommodate the pipes and their insulation.

Pumps in the standard range are somewhat too powerful for use with a small solar system and are not generally suitable for direct hot water supply circulation. Special 'solar' pumps have been

5.10   Cut-away view of solar pump

developed with the view to low power consumption and with characteristics suitable for handling small flows at a high frequency of switchings (up to 60 starts per hour). These pumps are now commercially available, see Fig. 5.9.

The velocity of the water in the pipes should be similar to central heating design criteria, ie, between 0.6 and 1.5 m/sec (2 and 5 ft/sec). At this velocity, heat is quickly removed from the solar collector and transferred to the heat store. If the pump is correctly sized to overcome pipework and equipment hydraulic resistance at

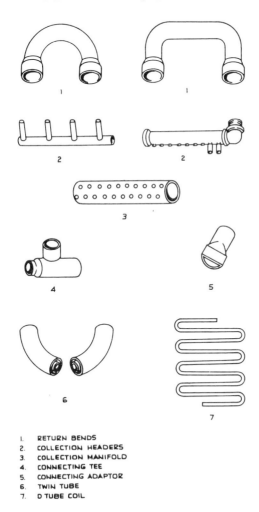

1.   RETURN BENDS
2.   COLLECTION HEADERS
3.   COLLECTION MANIFOLD
4.   CONNECTING TEE
5.   CONNECTING ADAPTOR
6.   TWIN TUBE
7.   D TUBE COIL

5.11   Standard fittings available in USA for solar installations

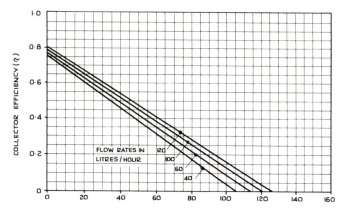

5.12   Graph showing typical relationship between flow rates through a collector panel and collector efficiency

the selected velocity, the electrical energy input to the pump motor is minimised – essential in any energy conservation exercise. Sizing of pipes is on the conventional basis, based on the above recommended velocity range. The total system resistance and required flow determine the pump performance requirement.

Alternatively, the installer may wish to select a particular pump and then design the system to suit the pump details.

For the more usual domestic installations, 75 mm (¾ in) pipe sizes will generally be found satisfactory.

Care must be taken in the selection of a pump for use with certain anti-freeze fluids and it is always advisable to check with the pump manufacturer on pump suitability.

The carefully chosen location and the correct installation of a pump is most important, particularly in roof spaces above the bedrooms, if complaints of noise and vibration are to be avoided.

The pump must be firmly supported and fitted with anti-vibration isolators and flexible pipeline connections. A well considered and engineered system should be reasonably free from noise or vibration.

The pipe connections onto the collectors should be arranged to allow flexibility of thermal movement to avoid stressing the collector(s).

As a sign of the rapid growth in the number of solar energy installations, a number of USA based manufacturers are supplying standard components for such installations. Fig. 5.11 indicates one such available selection.

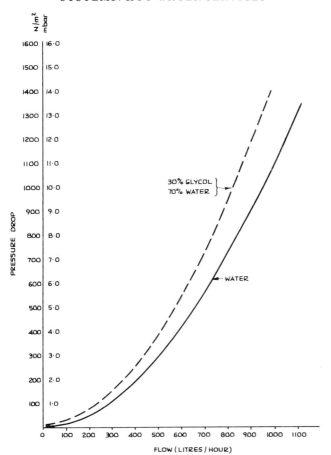

5.13   Graph showing typical relationship between flow rate through solar collector and pressure loss

*Reference:*
Institute of Heating and Ventilating Engineers, 1965, *Piped Services,* Section 12, page 262.

# 6

## SYSTEMS: CONTROLS

For optimum efficiency, solar heating processes must be controlled to operate without malfunction. Ideally, a solar hot water supply system should not operate unless solar energy is available. It should then operate effectively, stop when the demands have been met and protect itself against freezing and overheating.

Suitable control devices fall into two major categories:

    a)  self-acting (ie, without use of electricity)
    b)  electrically operated.

Self-operated controls are often preferred, as no auxiliary energy is used to drive the mechanism and the overall installation and operating costs are lower. The development of such devices for specific use in the solar industry has been slow and the availability of suitable controls is limited.

### Self-acting controls

Self-acting control systems do not require auxiliary power, such as electricity, compressed air or water to open or close the valves on actuation by a thermostat. They are widely used in space heating, hot water supply and process heating installations for the control of liquid and air temperatures and for weather compensating systems.

The typical self-acting water temperature controller comprises an immersion water thermostat containing a volatile liquid and a metal bellow, a control valve and a fixed length of capillary tubing connected between thermostat and valve. The temperature control operates on the principle of liquid expansion, the change in volume of the volatile expansion liquid on increase or decrease of temperature being utilised to open or shut the cone of the valve via metal bellows. Self-acting controls may be used in solar applications for example, to protect against overheating by using a primary fluid valve set to close at say 65.6°C (150°F) and a drain valve set to open a few degrees higher, so that the primary circuit is emptied automatically as overheat protection.

There are two major qualifications relating to this control arrangement:

the solar collector materials must be suitable for dry exposure to the sun
the drain water quantities on intermittent operation are not run to waste, but are returned into the system to comply with the generality of water companies' regulations.

### Electrically operated controls

These are more commonly applied to pumped systems. Each such control incorporates the following basic components.

*A measuring unit,* which responds directly to changes in the controlled conditions and provides a measured value accordingly; typical examples are the room thermostat, immersed duct thermostat, immersed water thermostat or externally located frost thermostat.

*A control unit* which adjusts the correcting condition in response to a signal from the automatic controller to vary the flow of the heating fluid, against a set control point. Typical correcting units are motorised valves and damper operators.

*An electric motor* which actuates the motorised valve, etc, and provides the necessary motive force for its movement. Motor capacities are small – in the range of 40–100 watt. An electric wiring system which links the various components and may include fuse protection, earthing, relays, air flow switches, contactors and the various other accessories to provide a complete and safe wiring installation.

*A centralised control point* at which the various controls are usually grouped together. This may take the form of an assembly of the various items of equipment fixed directly onto a suitable wall of an internal space or a neat purpose-made control box or panel. The control scheme may incorporate time controls, such as hand-wound time switches, synchronous electrically wound time switches with or without spring reserve, short period timers or changeover time switches. An electric time switch is usually provided. A spring reserve built into the time switch permits the switch movement to continue timing at times when the power supply is interrupted – a worthwhile and inexpensive ancillary, particularly in areas where interruptions to the electricity supply are expected to occur from time to time.

The most generally used device for a solar installation is a differential controller, which detects the temperature difference between the water in the collector panel and the water in the heat store.

In operation, the differential controller starts the circulating pump when the liquid in the solar collector reaches a temperature of about 5°C (9°F) higher than that in the heat store. The action of

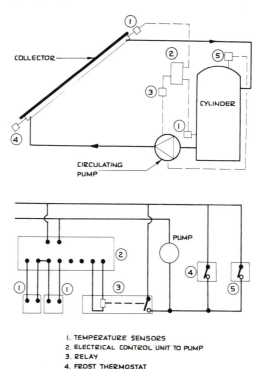

1. TEMPERATURE SENSORS
2. ELECTRICAL CONTROL UNIT TO PUMP
3. RELAY
4. FROST THERMOSTAT
5. HIGH LIMIT THERMOSTAT (ANTI- BOIL)

6.1    Solar heating circuit with frost protection by circulation the pump transfers the heat from the collector to the heat store (storage tank or cylinder). The controller stops the pump when the temperature differential has been reduced to a pre-set value. The differential is adjustable.

Where there is a back-up heating facility, by boiler or immersion heater, a simple automatic control protects the controller from freezing by the use of a frost thermostat. This switches on the pump frequently enough to transfer sufficient heat from the heat store to the collector – a somewhat wasteful method which can be obviated by the use of an anti-freeze solution (in an indirect system) or by the use of a frost protection thermostat which actuates valves which drain and vent the collector, but leave the storage cylinder isolated. See Fig. 6.1.

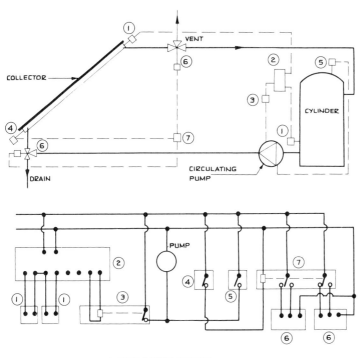

1. TEMPERATURE SENSORS
2. ELECTRICAL CONTROL UNIT TO PUMPS
3. RELAY
4. FROST THERMOSTAT
5. HIGH LIMIT THERMOSTAT (ANTI-BOIL)
6. 2 PORT MOTORIZED VALVE
7. RELAY

6.2   Solar heating circuit with frost protection by draining

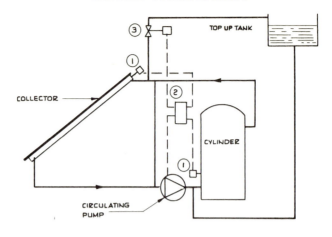

TOP UP TANK

COLLECTOR

CYLINDER

CIRCULATING
PUMP

— — — — ELECTRICAL WIRING

1. TEMPERATURE SENSORS
2. ELECTRICAL CONTROL UNIT TO PUMP
3. VENT VALVE

6.3  Control circuit with 'drying-out' frost protection

The correct positioning of temperature sensors is critical to successful operation and expert advice should be sought (say, from the control specialists).

There is the risk of boiling the water if little heat is drawn off during periods of high solar radiation. This can be overcome by an anti-boil thermostat which switches on the pump if the storage tank temperatures rise above 85°C (185°F) – irrespective of differential control. In this way, excess heat is radiated *from* the collector, until a safe temperature is reached.

A control system and the electrical connections for a solar heating circuit with frost protection by draining is described in Fig. 6.2. A further scheme, shown in Fig. 6.3, provides the necessary differential temperature control for drawing heat from the collector and also incorporates a further alternative for frost protection. In this case, the collector is dried out during frost conditions, simply by the control action of running the pump whilst the vent valve is open – the valve then closes, leaving an air lock in the collector and the pump stops. When the temperature rises above freezing, the vent valve opens without the pump running and the collector refills, after which the valve closes, returning the heating circuit to its normal condition.

Another overheat protective arrangement incorporates a fan-convector (or fan assembly and a car-type radiator) connected by

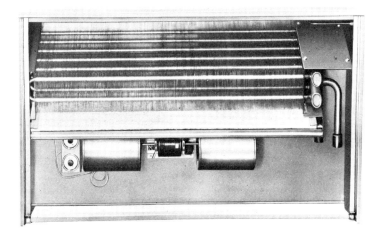

6.4    Fan-assisted heat exchanger for spillage of surplus solar heat (front cover removed)

pipes to a supplementary primary flow pipe to cool the overheated hot water by dissipating the unwanted heat into the roof space or to outside, the fan being switched by an overheat or limit temperature sensor. In certain applications, this surplus heat can be used effectively, eg, for clothes or grain drying. Fig. 6.4 illustrates such a convector suitable for a larger system.

There is also an overheat device available which includes motorised louvres fitted above, or just within, the solar collector. The louvres close to terminate heat collection, actuated by temperature sensors. This control offers the advantage of protecting also the material of the solar collector, but it tends to be costly.

The purchaser of solar equipment should look out for relatively simple and straightforward control systems, perhaps with the refinement of a visual operating indication. An enthusiastic tendency towards over-specification should be resisted, as over-elaborate control systems might cost almost as much as the installation itself, thereby defeating the aim of overall energy conservation.

Temperature sensors should not only be carefully located to sense (detect) the conditions they are to control, but they must also be protected in their working environment and be easily accessible for routine maintenance and replacement.

# 7

## SYSTEMS:
## OTHER THAN HOT WATER SERVICES

In general, solar heat reception and space heating requirements do not coincide. A further drawback to solar space heating lies in the often only intermittent reception of solar heat. Nevertheless, there are numerous situations in which solar space heating can prove profitable and worthwhile.

It is easily understood that the complication of intermittance can be met by adequate storage facilities to spread the benefit of the solar reception over those periods of the day during which there is no reception of solar heat.

There are many locations where there is a rapid drop in the air temperature after sunset, following intense sunshine during the day. Such circumstances offer the opportunity for providing space heating during the cool evenings and nights derived from solar heat received during the day.

A correctly designed heat store is required to carry over the benefit of solar heat to the periods when space heating is desired.

While for heat storage purposes water is the best material on a volume basis, solids such as crushed rock, concrete or brick, scrap metal and magnetite may, for certain applications, prove to be economically superior. The primary considerations are the ultimate use of the energy and the means of transporting (conveying) it. The main methods of storing solar heat currently in use are:

**Water storage tanks**
To be effective all day round, such tanks require to be of adequate

size to deal with the 24-hour requirement for hot water. They must be efficiently insulated to maintain the temperature of the stored water during the storage period. As a back-up heat source, one, or several, electric immersion heaters are commonly inserted in the storage tank – limited to areas served with electricity.

Alternatively, the back-up heat may be provided by a primary coil in the heat store through which boiler heated hot water is circulated.

**Pebble beds**

Pebble beds, or rock piles, can be used for storing heat at a relatively high temperature. Whilst the lateral conduction of heat from the pebbles is low because of the small contact areas, the exchange of heat between the moving carrier air stream and a pebble bed is effective and quick acting.

Ordinary rock or stone has a fairly high thermal capacity. In pebble-bed storage, washed stones of 19 mm to 32 mm (¾ in to 1¼ in) size are best used. In some heat store beds oil has been mixed with rock to improve the heat transfer. There is currently experimentation with various combinations to improve thermal capacity, reduce the size of heat store and provide a reliable source of heating supplement. The technology is not new – Electricity Boards and Utility Companies have used rock-type night store space heaters for many years.

Since the air drawn across the heat store will be supplied into the user premises, it is essential to ensure that the materials of the heat store are clean and free from objectionable smells. The principle of the rock type heat store may be extended to using the structure of the building itself as heat store or to a method of locating the rock between two leaves of a cavity wall.

**Heat-of-fusion heat storage**

This method utilises heat storage materials which melt at a moderate temperature level and store heat as their heat-of-fusion, or heat of transition. Low-cost salt hydrates, which are easily available – some of them being obtainable as by-products – can be used for this purpose. A typical example is sodium sulphate decahydrate ($Na_2SO_4$ $10H_2O$); this almost entirely melts in its water of crystallisation when heated to its transition temperature at 32.2°C (90°F). Its heat-of-fusion is 242 kJ/kg (104 Btu/lb) and the density is 147 kg/m$^3$ (92 lb/ft$^3$). Therefore, one cubic metre can store 345 MJ (one cubic foot of this material can store 9,500 Btu) as its heat-of-fusion at the transition temperature.

The specific heat of this material is comparable to that of water on an equal volume basis; the stored heat can be recovered as the material crystallises again. With a temperature rise of 18°C (32.4°F), the total heat storage capacity of this material can be as high as 443 MJ/m³ (11,900 Btu per ft³). This is six times greater than that of water and eleven times greater than that of rocks, on an equal-volume basis (when other temperature intervals are used, these values may be different). Thus, with the heat-of-fusion type materials, it is possible to store more heat per unit volume or, conversely, to use a smaller heat storage volume relative to water or rock storage.

The cost of these materials may be quite low, but as materials of this type change from the solid to the liquid state and, therefore, must be stored in suitable containers, some extra costs may be involved. Table 7.1 lists the properties of several salt hydrates which are considered suitable as heat-of-fusion materials. The transition temperature varies according to the formation of salt hydrates containing fewer water molecules, and the heat-of-fusion may change accordingly.

The process of fusion is theoretically reversible and the entire heat input should therefore be recoverable. There are, though, several limitations, which may delay the prompt recovery of the stored heat: when the materials are melted completely in sealed containers, they may not solidify on cooling, but undercool below their melting points; (undercooling can be prevented by using crystallisation catalysts or nucleating agents); the other limiting factor is the rate of crystal growth which limits the rate of heat exchange between the solid and liquid phases. Most of the listed salt hydrates have a limited crystallisation velocity of the order of 0.89 mm (0.035 in) per hour per °C temperature difference between solid and liquid. Thus, heat cannot be withdrawn more rapidly than it can be supplied by the growth of the the crystals; for practical purposes, the formation of crystals does not progress more rapidly than at the rate of 12.7 mm (0.5 in) per day. It is necessary to provide a suffcently extensive heat transfer area between the heat transfer fluid (air) and the heat storage material, which is sealed into containers of suitable geometrical shape.

Reference to Fig. 7.1 shows the manner of operation of a rock-store space heating system. During the day, air is circulated through the solar collector and is warmed; it is then conveyed by the 'day' fan into the rockstore which heats up progressively dur-

**Table 7.1    Thermal properties of heat-of-fusion storage materials**

| Compound | Transition °C temperature °F | | kj/kg | Heat-of-fusion Btu/lb |
|---|---|---|---|---|
| $CaCl_2$ $6H_2O$ | 29–39 | 84–102 | 174.5 | 75 |
| $Na_2CO_3$ $10H_2O$ | 32–36 | 90–97 | 267.5 | 115 |
| $Na_2HPO_4$ $12H_2O$ | 36 | 97 | 265.2 | 114 |
| $CA(NO_a)_2$ $4H_2O$ | 40–42 | 104–108 | 209.3 | 90 |
| $Na_2SO_4$ $10H_2O$ | 32 | 90 | 242.0 | 104 |
| $Na_2S_2O_3$ $5H_2O$ | 49–51 | 120–124 | 209.3 | 90 |

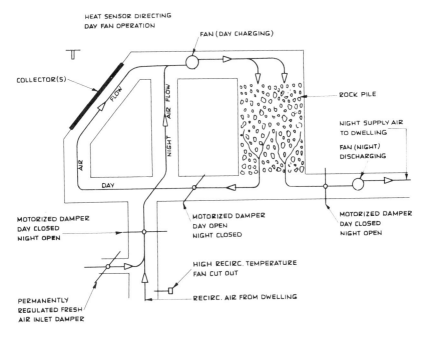

7.1 Solar warm air heating system with rock store

ing the day. Once the external air temperature has fallen to defeat further solar collection, the 'day' fan is switched off and the controls close down the circuit between the collector and the heat store. When the thermostat in the user premises calls for space heating, the 'night' fan starts up, dampers are automatically set and the warm air is drawn into the premises by the continuous circulation of air between the store and the premises. Some controlled intake of fresh air is generally desirable for odour control, but this should be adjusted for a minimum flow of air to conserve

**Table 7.2   Recommended heat store/energy input relations**

| Building energy input average requirement | Minimum heat store capacity in litres (gallons) | | | | |
|---|---|---|---|---|---|
| | 10 hrs storage | 20 hrs storage | 30 hrs storage | 40 hrs storage | 50 hrs storage |
| Multiplier | 1.0 | 2.1 | 3.2 | 4.3 | 5.4 |
| kW (Btu/hr) | | | | | |
| 14.7 | 4,546 | 9,547 | 14,547 | 19,548 | 24,548 |
| (50,000) | (1,000) | (2,200) | (3,200) | (4,300) | (5,400) |
| 29.3 | 9,092 | 19,093 | 28,185 | 39,096 | 49,097 |
| (100,000) | (2,000) | (4,400) | (6,400) | (8,600) | (10,800) |
| 44 | 13,638 | 28,640 | 42,278 | 58,643 | 73,645 |
| (150,000) | (3,000) | (6,300) | (9,600) | (12,900) | (16,200) |
| 58.6 | 18,184 | 38,186 | 56,370 | 72,191 | 98,194 |
| (200,000) | (4,000) | (8,400) | (2,800) | (17,200) | (21,600) |
| 73.3 | 22,730 | 47,733 | 70,463 | 97,737 | 122,742 |
| (250,000) | (5,000) | (10,500) | (16,000) | (21,500) | (27,000) |
| 88 | 27,276 | 57,280 | 84,556 | 117,287 | 147,290 |
| (300,000) | (6,000) | (12,600) | (19,200) | (25,800) | (32,400) |
| 102.6 | 31,822 | 66,826 | 98,648 | 136,835 | 171,839 |
| (350,000) | (7,000) | (14,700) | (22,400) | (30,100) | (37,800) |
| 117.2 | 36,368 | 76,373 | 112,740 | 156,382 | 196,387 |
| (400,000) | (8,000) | (16,800) | (25,600) | (34,400) | (43,200) |
| 132 | 40,914 | 85,919 | 126,833 | 175,930 | 220,936 |
| (450,000) | (9,000) | (18,900) | (28,800) | (38,700) | (48,600) |
| 146.6 | 45,460 | 95,466 | 140,926 | 195,478 | 245,484 |
| (500,000) | (10,000) | (21,000) | (32,000) | (43,000) | (54,000) |

Notes: The table assumes a minimum thermal insulation efficiency of 90%

For *other materials*:

multiply by $\dfrac{\text{inverse ratio of specific heats}}{2,240} \times 10$ tons

(1 ton (imperial) = 1,016 kg)

the stored heat. Next day, the solar collector is brought back into use, with the heat being directed into the rock store in readiness for the eventual space heating requirement.

There may also be circumstances when there is a clear sky on a cold day and space heating is desired in daytime. The controls and fans can be adapted to meet such a situation. However, the system shown in Fig. 7.1 is arranged for only night time space heating. Where solar heated domestic hot water is also required, it is best to separate the two solar functions and to have a separate collector and system for the hot water supply.

### Relationship of heat store to system

Many variables determine the size of heat store and the associated system; Table 7.2, indicates recommended sizes of heat store, related to energy input. It will be noted that relatively large heat storage capacities are required, even for the smaller systems.

The principle of the rock type heat store may be extended to using the structure of the building itself as heat store or to locating the rock between two leaves of a cavity wall.

HEATING SWIMMING POOLS BY SOLAR ENERGY

The most obvious and direct method of heating an outdoor swimming pool is through the direct absorption by the water of the incident solar radiation; in a temperate climate, such as in the UK, the temperature of an unprotected and unshielded pool would closely follow the mean air temperature over the summer months.

The uncertainty of the weather conditions and the common desire of swimming pool owners to extend the use of the pool to spring and autumn (or in warmer climates, throughout the year), generally used to involve the adoption of a conventionally fuelled heating system – usually by gas, oil or electricity (or where available, off-peak electricity). Since the costs of such fuels are proving more and more expensive, solar heating has gained in popularity and many solar heated swimming pools have been constructed and critically observed. The major specific factors to be considered when designing the solar heating arrangements for an outdoor swimming pool are:

Reduction of the heat losses from the pool water
Solar heating equipment
Provision of back-up heating, in addition to the solar heaters.

## Collectors for swimming pools

The type of solar collector previously discussed for use with hot water and space heating services has to raise the temperature of the water adequately to provide useable hot water without the need for constant back-up heating. The minimum temperature required for such application would be about 52°C (126°F). The solar collectors are designed accordingly and hence are generally provided with a double glazed cover to permit high efficiency collection and trapping of heat.

The collectors used in swimming pools are inserted into the pipe circuit between the pool and the pool water filter. The temperature requirements relate to that to be maintained in the swimming pool, which seldom need be more than 26.7°C (80°F). In fact, a lower temperature is quite often acceptable.

It is thus evident that the difference in temperature ($\Delta T$) between the pool of heated water and the collector can be less than for the conventional solar panel; this is reflected in the design of the swimming pool collector. It is not necessary to provide double glazing; unglazed collectors are suitable. The construction cannot be of copper because of chemical attack by the chlorine. Plastic construction is employed for this purpose and must be compatible with the materials used in the swimming pool plumbing.

The collectors need not necessarily be located on a roof above the swiming pool; they may be fixed on to frames in ground positions, so long as the collectors have an unobstructed access to sunshine. Some experimentation has also been carried out with the collectors built into the steps and terraces surrounding the swimming pool. The collector must be of the pressure type.

Fig. 7.2 indicates a typical circuit for a swimming pool solar system.

## Reduction of heat losses

A swimming pool loses heat through evaporation, convection and radiation. The conduction losses are usually relatively minor and can be neglected.

Where the pool is wholly submerged in the ground, nearly all the heat that escapes into the pool returns to it on a drop in pool water temperature. If the pool is not fully excavated, a heat loss occurs from the exposed wall sections.

The out-of-use evaporative loss can be almost completely eliminated by the use of a single thin cover placed over the water surface; recent developments in plastics have greatly increased the

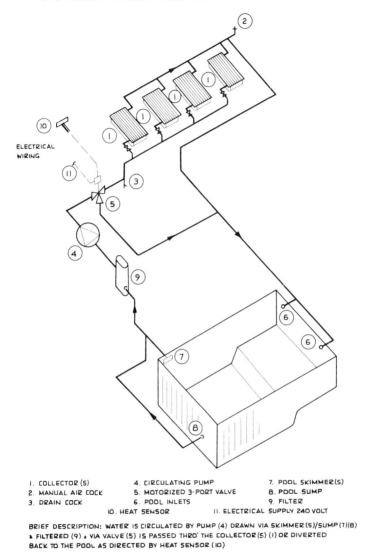

ELECTRICAL
WIRING

| 1. COLLECTOR (S) | 4. CIRCULATING PUMP | 7. POOL SKIMMER(S) |
|---|---|---|
| 2. MANUAL AIR COCK | 5. MOTORIZED 3-PORT VALVE | 8. POOL SUMP |
| 3. DRAIN COCK | 6. POOL INLETS | 9. FILTER |
| | 10. HEAT SENSOR | 11. ELECTRICAL SUPPLY 240 VOLT |

BRIEF DESCRIPTION: WATER IS CIRCULATED BY PUMP (4) DRAWN VIA SKIMMER (S)/SUMP (7)(8)
& FILTERED (9) & VIA VALVE (5) IS PASSED THRO' THE COLLECTOR (S) (1) OR DIVERTED
BACK TO THE POOL AS DIRECTED BY HEAT SENSOR (10)

7.2 Schematic arrangement of solar-heated swimming pool

range and practical applications of the various easily-handled types
of swimming pool covers.

Covers may be a double-layer plastic sheet with intermediate air
space: a simple light gauge black polythene sheet laid on the pool
surface and clamped around the edges of the pool; a double layer
of blue PVC sheet separated by strips of polyurethane foam.

A description of the 'Bahama Pool' solar heated swimming pool at Ye Old Felbridge Hotel, East Grinstead, England, is of interest. The solar heat collector has an effective heat collecting area of approximately 50 m² (538 ft²) and was designed for a maximum heat collection rate of 755 W/m² (239 Btu/hr/ft²), when operating at a temperature of 25°C (77°F). Originally painted matt black to increase the solar absorption efficiency, it was repainted light green to overcome complaints of the intense reflection of the morning sun into some of the hotel bedrooms.

To maintain efficiency, the collector is occasionally hosed down; no further work is required on the system during the season. No pool cover is used because of the practical difficulties of manipulating this over a much-used large pool. It has been established that, although the mean temperature in the pool during the day is some 3–6°C (5.4–10.8°F), greater than that of a similar non-heated pool in the same district, the nightly drop in temperature is usually no greater than 1°C (1.8°F).

The solar collector is inclined at 60° to the horizontal and faces 10° west of south, to achieve a greater efficiency in the late afternoon and towards the end of the season.

The design intent was to maintain the pool temperature at 22°C (72°F) during the months of June to August; this aim has been achieved. A back-up electric booster heater was subsequently added to ensure that swimming can continue to be enjoyed in the late spring and early autumn when the sun effect is uncertain.

Photograph 7.3 shows a 'Solaris', highly efficient low-cost solar collector, especially designed for the heating of swimming pools. Unlike the smaller conventional collectors, with their multiple arrays and complex inter-connecting pipework, the Solaris is a large unitary collector with two main manifolds (headers) for water supply and return.

The collectors come in kit form and consist of a patented reflecting surface which concentrates the solar radiation, together with black tubes which absorb the radiation and thereby heat the water which circulates through the tubes. The tubes themselves are manufactured of ultra-violet resistant high-conduction plastic materials and they are clipped into the reflecting surface.

The tubes are jointed with the manifolds by compression type connections. These complete panels may be connected in series with the circulating and filtration system of the swimming pool. The collectors are in widths of 2 m (6.6 ft) and supplied in three different lengths of 12 m (39.4 ft). 16 m (52.5 ft) and 20 m

7.3 A 'Solaris' collector in Tel-Aviv, Israel

(65.6 ft) respectively. The manufacturers also provide associated transparent covers.

A rule of thumb for the determination of the number of collectors required for a particular application is to calculate the surface area of the pool and employ a collector area equal to a *minimum* of half that area. Experience indicates that the most effective collector area is 75% of the pool area, while the technically most efficient solution calls for 100%.

The recommended flow rates for the Solaris panels are 4,200 litres/hour (925 gall/hr) for a 40 m² (430 ft²) collector, 3,400 litres/hr (749 gall/hr) for a 32 m² (345 ft²) collector and 2,600 litres/hr (573 gall/hr) for a 24 m² (258 ft²) collector.

## SOLAR DRYING SYSTEMS

The drying of produce is a solar energy application which has received only limited and spasmodic attention over recent years, although it appears to offer worthwhile benefits.

The drying process and system applications incorporating a solar supplement must be well thought out.

Sun-drying of washing, hung on a clothes-line, appears to be simple, but conventionally fuelled timber drying kilns, for instance, operate under strictly controlled conditions to provide a uniform product of pre-selected quality.

In areas of Australia, timber drying with a solar supplement is now well tried and tested and constitutes a normal approach to drying. Timber driers attempt to follow a set schedule of achieving and maintaining kiln wet and dry bulb temperatures without resorting to auxiliary heat input, in conjunction with a heat store formed of gravel; provision is made for air flow either upwards or downwards through the bed. When available, solar heat is used to maintain the required temperatures; when there is excess heat, it is diverted to the heat store for subsequent use. At night time and at times when the sun is not shining, reverse flow through the gravel bed provides heat for a continuous process of drying.

Grain drying has not been given much similar attention, possibly because of grain hardening problems and the semi-permanent nature of the stored produce. There is clearly considerable potential for solar processes of this kind.

### GREENHOUSE HEATING

Greenhouses are effective solar collectors. The obvious need in terms of effective comprehensive solar applications is to integrate the thermal storage of a heat store into the greenhouse assembly and to divert the surplus heat into the heat store for use during periods when there is no sunshine.

An entirely different greenhouse construction concept might have to be adopted to facilitate the practical exploitation of solar energy. Practicable possibilities are the provision of suspended rotating drum type, change-of-state storage units and below-ground gravel, rock or earth stores.

### INDUSTRIAL AND COMMERCIAL PROCESSES

Many industrial and commercial processes require large quantities of hot water. Milking parlours, breweries, cleansing and washing industries are some of the many heat users which could adapt existing water storage equipment to benefit from solar energy reception, particularly where there is ample space for the location of the collectors and heat stores.

Steam boiler houses usually contain large-size treated or raw feed water tanks. Heating part of the sensible heat requirement by solar means can bring substantial savings.

### CAMPING – SOLAR APPLICATIONS

This is an obvious area for a solar application, where there is a frequent need for hot water and there is, in many cases, no means of heating this except by solar energy. Portable solar collectors for campers' use are now offered by a number of manufacturers.

A major problem with portable camping units is that by, their very nature, they are small and therefore able to collect only limited amounts of solar energy; this is likely to render the appliance of limited use and, at times, impracticable. Reflectors can be used and are often provided, but the difficulty of manually adjusting them to correct the variable sun angles does present a problem.

It has been shown that relatively large areas of collector surface are required for practical purposes and this opens up the possibility of collector placement on top of caravan roofs. Caravan roofs can be used for hot water supply collection, with perhaps a central 'dome' heat store; they also afford a distinctly attractive collecting surface for integrated heating and cooling solar appliances because of the relative ease of correct orientation and limited thermal loads.

### USER HABITS

Over a period of time, the occupancy and habits of a household change: the family grows up; the house changes ownership; there may be other domestic reasons. It is, therefore, advisable to relate the size of a proposed solar energy installation to the probable occupancy and probable user behaviour.

The design relationship given in chapter 4 between occupancy and collector size is essentially a guide and obviously cannot apply to all cases. The requirement of a family with many young children, which involves much hot water use during the day time, is different to that of an equally numerous family whose members are out at work all day. In the latter case, a larger heat store might well be desirable to store as much hot water as is practically possible for use during the evenings, when the maximum demand will arise.

**Table 7.3   Solar energy system based on user habits**

| Occupancy (no of persons) | Behaviour | Average family water usage per day litres (gallons) | Recommended heat store capacity litres (gallons) | Recommended size of collector array $m^2$ ($ft^2$) | May–August probable utilisation factor | Remaining 6 months probable user factor |
|---|---|---|---|---|---|---|
| 2 | Both at work | 136 (30) | 136 (30) | 3 (32.3) | 0.386 | 0.772 |
| 2 | Wife at home | 341 (75) | 136 (30) | 3 (32.3) | 0.9 | 0.95 |
| 3 | Both at work | 341 (75) | 136 (30) | 4 (43.1) | 0.257 | 0.514 |
| 3 | Wife at home | 455 (100) | 136 (30) | 4 (43.1) | 0.9 | 0.95 |
| 4 | Both at work | 455 (100) | 272 (60) | 5 (53.8) | 0.386 | 0.772 |
| 4 | Wife at home | 568 (125) | 272 (60) | 5 (53.8) | 0.9 | 0.95 |
| 5 | Both at work | 568 (125) | 272 (60) | 6 (64.6) | 0.31 | 0.61 |
| 5 | Wife at home | 796 (175) | 272 (60) | 6 (64.6) | 0.9 | 0.95 |
| 6 | Both at work | 682 (150) | 272 (60) | 8 (86.1) | 0.257 | 0.514 |
| 6 | Wife at home | 910 (200) | 272 (60) | 8 (86.1) | 0.81 | 0.855 |

Notes:   The table relates to average middle-class families with appropriate occupations, hobbies and habits. If some members of the families are engaged in more dirty work, such as garage mechanic, maintenance or workshop fitter, building site operative etc, then the nature of hobbies should be considered relative to additional washing and washing-up requirements. To compensate, add one-third to total average family probable water usage per day. User factors may then be re-calculated in accordance with text.

Under poor insolation conditions, utilisation may well be unity under any circumstances. If heat-store is under-sized, then utilisation must be poor, unless hot water is used frequently during day-time. Over-night thermal losses from heat store must be considered for early morning ablution. It is probable that families with solar systems tend to use more hot water.

Unless water meters are fitted to record how much hot water is used by different categories of house occupiers, it is necessary to have some knowledge of 'user habit', if a realistic stab is to be made at the assessment of actual saving due to arise from solar energy use. Some guidance is given below.

Consider a two-person family, both persons being out at work five days each week; during the remaining two days they are at home, cleaning the house, doing the week's washing, etc.

During the two days at home each week, it may be assumed that there is full utilisation of the hot water facility, but during the other five days, perhaps only half utilisation. A 'user factor' may then be easily calculated as the ratio of 4.5:7 or 0.643 (the numerator 4.5 arises as to: 5 day/2 plus 2 days/1).

This calculation may not be entirely accurate, as during the sunny summer the collector array may have the ability to reheat the heat store three times per day, whereas in the spring and the autumn less than twice per day, on average. There is also a practical limit as to the quantity of hot water a person is likely to use in one day for ablution purposes; in the industrialised countries the consumption is in the order of 57 to 114 litres (15 to 30 gallons) per person. Many other factors enter into the assessment, eg, occupation (clean or dirty job); low, middle or high income group; leisure time; sports activities; hobbies, climate, etc: these considerations make it very difficult to accurately assess the 'user factor'.

Table 7.3 has been prepared to serve as a guide. It is intended to be used with discretion and consideration of all factors pertaining to likely use and latitude of insolation.

EXAMPLES OF WORKING SOLAR INSTALLATIONS

Photograph 7.4 shows an extensive operative solar energy installation serving an automatic service station in Goldfield, USA, a location in the high Nevada desert at an altitude of 1,737 m (5,700 ft). The hours of sunshine actually experienced since commissioning the plant appear to be in excess of 80% of the possible sunshine hours.

Three independent arrays of collectors are installed. Any given collector can be worked on without turning off the array. There are 49 dual-glazed collectors each of dimension 0.91 m×2.44 m (36×96 in) nominal and having 2.1 m$^2$ (22.65 ft$^2$) nett glazing area (as the aperture). Glass is 4.76 mm ($^3/_{16}$ in) low iron tempered type, spaced 12.7 mm (½ in). The absorber is stainless steel

7.4  Solar installation at Goldfield, USA

with nickel flash and coated with electroplated black chrome. The arrays each consist of 14, 17 and 18 collectors, respectively. The collectors have been incorporated into the waterproof membrane of the roofing (this expense cannot always be justified as it is intensive in site labour).

The orientation of the arrays is to the true south, inclined approximately 54° from the horizontal.

It was considered that the epoxy or special paints could not be successfully used on double-glazed collectors because they cannot stand the higher temperatures without smoking the glass. (Good epoxy paint can be used in single-glazed collectors successfully, but extra steps have to be taken to protect from stagnation should the pump fail). The worst possible circumstances must be considered until more is known or the paints are improved. In the meantime and if there is doubt, selective coatings should be used. Before the actual construction, the absorber coating specification was changed from epoxy paint to black chrome.

The flow rate has been varied to compare output temperatures; in use, a water temperature of 93.3 °C (200 °F) has been easily obtained.

The system employs a 50% ethylene glycol-water mixture in the collector zones and transfers the heat via heat exchangers in the storage tanks to the interior circulating heating water system.

The installation takes care of about three-quarters of the annual heating costs for the building and of virtually all the small domestic hot water requirements. Auxiliary back-up is by an oil-fired hot water boiler.

The above system has been operational for some time. There have been very few problems; intermittent instrument checks have proved that the system will meet at least the original design expectations.

A most interesting and advanced solar collection project is reported from Maherst, Massachusetts, USA. The architects for the five-building Arts Complex at the Arts Village of Hampshire College wanted a solar system to serve the entire complex, despite a phased construction schedule. It was their aim to incorporate the solar system as a major design feature of the complex in order to architecturally relate the collectors to the buildings. To achieve the design intention, rows of all-glass, selectively coated evacuated tubular collectors were transformed into a string of canopies on a bright-blue structure between the buildings. The result: a functional and clean environmental sculpture overhead and sheltered building walkways below. This design solution moved the collectors from the building roofs and thereby eliminated the considerable loads which would have been imposed by the collectors, so that simple buildings of pre-fabricated components could be constructed without the constraint of the collectors.

The Arts Village solar system was funded by a US Federal grant of US$355,000. It is anticipated that the system will supply 95% of the cooling needs, 65% of the space heating and 100% of the domestic hot water requirements of the complex. The co-ordination of the solar system is by a microprocessor and the back-up heating from a central electric boiler plant.

Another application of solar energy may well be adopted widely in the developing countries: it is a solar-oven originated by Father Julien, a Jesuit Missionary in Chad, an African country located on the southern edge of the Sahara desert, at a time when prolonged periods of drought had caused serious and widespread damage to the tree plantations, so essential for a wood-burning economy. The technical problems of the early design, which worked well for the villagers' needs, but was inadequate for the brickmakers and blacksmiths, were taken over by the École Catholique d'Arts et

Métiers (ECAM) in Lyon, France and backed by the aid organisation Secours Catholique.

ECAM built a prototype brick kiln of 11 kW output to work at a maximum temperature of 1,000°C (1,832°F). It had to be cheap, simple to install and to maintain, while achieving an output of 50 kg of clay per day. To concentrate the sun's rays, two heliostats (essentially movable mirrors) were employed, one rotating on a perpendicular axis which is daily adjusted. The system has proved itself in use; similar solar powered ovens can be used for iron and aluminium, light alloys, forges, kilns and for smaller domestic heating and cooking applications.

The spread of such solar-powered devices throughout Africa will confer the benefits of freely available solar energy and will greatly encourage the planting and preservation of the trees which are vital barriers in the control of soil erosion.

Photographs 7.5, 7.6 and 7.7 show the solar heated swimming pool at the School of S. Mary and S. Anne, Abbots Bromley, Staffordshire. The pool is fully air-conditioned and it was decided to

7.5

7.6

7.7

incorporate a solar heating system. The selection of a suitable solar panel was complicated by the various claims and differing information given by the manufacturers. More than twelve companies were contacted by the architects and a short list was made of those who impressed with their efficiency and technical knowledge.

The type of panels offered were:

a)  Polypropylene panels
b)  Metal collectors with glazing.

There was a considerable difference of opinion as to the total area of panel required for the solar heating to the pool. However, there was no doubt that polypropylene panels were not self-cleansing and would probably have a limited life of some ten years (although there is as yet inadequate in-use information to confirm this view).

It was decided to adopt EAS glazed metal collectors, each comprising a collecting plate of copper tubing with collector fins. The whole panel was enclosed in a fully glazed and insulated galvanised frame. The total installation consists of 63 panels, each a Silentair EC2 panel, 1.07 m $\times$ 2.0 m (3.3 ft $\times$ 6.6 ft), giving a total collector area of 134m² (1,442 ft²).

It is anticipated that the installation should save approximately 83,000 kW (28.5 million Btu) of useful heat per annum, which could equate to an annual saving in running costs of £600 to £700 (US$1,200 to 1,400).

The solar installation has been arranged on the 'direct system'. The operation is controlled by differential thermostats. Frost protection is by a frost thermostat actuated by drain-down pipe and control circuit.

*References:*

'High performance solar collectors form canopied walk at experimental college arts complex', *Architectural Record*, mid-August 1978, page 94

'Solar Energy and Soil Erosion', *Building Design*, October 20, 1978, page 44

# 8

## ENERGY CONSERVATION

The object of installing a solar energy system is to save money by displacing more conventional and expensive forms of energy. Thus, one should not look in isolation at the potential savings due to arise from solar energy systems. One ought to first consider whether the financial allocation to a solar energy system could be more beneficially expended in eliminating the wasteful use of energy within the premises, so that the *overall* energy picture can be evaluated. To facilitate such study, an energy conservation check list is set out below.

ENERGY CONSERVATION – CHECK LIST

Note: this list applies to residential, commercial and factory premises

1. **Building construction**

    *Check*
    1.1    That the ceilings and roof are thermally insulated to adequate standard
    1.2    On insulation of walls – could cavity fill be economically justified?
    1.3    On weather and wind tightness of all outside doors and of windows – much heat can be lost through excessive cracks and gaps

1.4    On economic feasibility of double glazing all windows, par-
       ticularly of those facing north
1.5    In high warehouse/factory bays, on the temperature of the
       air at the underside of the ceiling/roof. If excessively high,
       there will be a high heat loss through the roof. Consider
       feasibility of providing a lower suspended ceiling or of cir-
       culating the air from high level downwards.


2.     **Boiler plants**

       *Check*
2.1    That the correct boiler fuel is in use
2.2    Combustion efficiency – set plant to operate with
       minimum excess air and smoke number
2.3    That all combustion components, such as nozzles, refrac-
       tory and boiler doors are in prime condition; replace as
       necessary
2.4    That the boiler tubes/flue-ways and chimneys are free from
       soot and cleaned adequately at regular intervals
2.5    On sequence operation of multiple boiler installations – do
       not operate boilers on partial load, which could be serviced
       from one (or a smaller number) of boiler units
2.6    That the system is temperature or heat timer controlled to
       minimise operating waste
2.7    Regular servicing arrangements for temperature controls to
       maintain same in constant good order
2.8    That the hot water supply is not hotter than required
2.9    That steam boiler plants do not operate with excessive
       blowdown and/or leaking or blowing safety valves
2.10   That all hot surfaces and hot tanks are thermally insulated
       to an adequate standard
2.11   That ventilation to boiler room is adequate to obviate
       excessive ambient temperatures which might damage con-
       trol equipment
2.12   That the glands on circulating and feed pumps are in good
       order
2.13   That programme and time controls are incorporated and
       are set to meet the correct requirements
2.14   That efficient frost protection is provided and that the
       associated thermostats are correctly set
2.15   That water treatment/conditioning is efficient and super-
       vised.

3.    **Furnaces and ovens**

*Check*
3.1   On the feasibility of employing a higher standard of heat insulating material
3.2   That furnaces work for minimum operation time; they are often run partly loaded all day, whereas planning might enable the same work to be done in half a day with the furnace fully loaded
3.3   Whether the mass of carriers, trays and other furniture in an oven could be reduced
3.4   That furnace doors are left open for loading and unloading operations for the shortest possible time.

4.    **Ventilation and air conditioning**

*Check*
4.1   That there is no excessive fresh air ventilation
4.2   That the windows are normally closed in cold weather; if not, this indicates that the temperature is too high or the ventilation is not working properly
4.3   That fans are regularly cleaned and lubricated
4.4   That the air ducting is regularly inspected for leaks and that these are sealed where necessary
4.5   That the controls on air-conditioned buildings are effective and maintain the correct temperature everywhere; over-heating/excessive cooling is costly
4.6   Where it is practicable/safe to shut down ventilation one hour before work stops
4.7   Whether there is any exhaust heat that could be used for drying or any other part of the process
4.8   Whether the discharge of large quantities of warmed exhaust air justifies the provision of heat recovery systems, such as heat wheels, heat pumps, wrap-around heat exchangers, etc.

5.    **Space heating**

*Check*
5.1   Whether it is practicable to reduce heating of unoccupied areas
5.2   The accuracy of space heating thermostatic control regularly

5.3     That doorways are kept closed when heating is on by fitting springs, etc

5.4     That factory/warehouse doorways in constant use are fitted with rubber doors or similar protection

5.5     That the warm air curtains on doors are only used when doors are open and on the coldest days

5.6     That the temperatures are carefully regulated relative to use of premises

5.7     That heating at night and on non-working days is reduced to a level just sufficient to give frost protection

5.8     That any temporary heating methods (particularly in offices) have not become permanent

5.9     That fan heaters are properly maintained and radiators regularly cleaned

5.10    Whether draught lobbies are practicable at external doors

5.11    That all hot pipes are thermally insulated in locations where they do not contribute to controlled space heating

5.12    That all underground heating pipes are adequately thermally and corrosion insulated.

## 6.    Hot water and steam transmission

*Check*

6.1     That all lagging is in good order; consider feasibility of increasing the thickness of thermal insulation

6.2     That all redundant pipework is removed or isolated

6.3     That all steam traps are subject to inspection and maintenance

6.4     That all heated pipes not required for space heating are lagged

6.5     That all steam, condensate and water leakage is repaired

6.6     That there is maximum percentage return of condensate to boiler plant

6.7     That all automatic/temperature controls are in good order

6.8     That all automatic air release valves are in good order

6.9     That all hot water blenders are correctly set and in good order

6.10    That 'economy' type hot water taps and showers are fitted wherever practicable.

## 7.    Compressed air service

*Check*

7.1     Thoroughly and regularly for leaks and repair where necessary; a small leak can represent a considerable loss of energy

7.2   The system regularly for redundant parts and seal them off
7.3   That compressor efficiency is maintained by renewing valves, rings glands, etc
7.4   That filters are clean and renewed regularly
7.5   That the compressor is switched off when not required, and sometime before work ceases, so that the reservoir is not left full of compressed air to leak overnight
7.6   That the water is regularly drained from the system
7.7   Whether heat can be reclaimed from the compressor and receiver arrangement, such as for use in heating the hot water supply to kitchens, toilets, etc.

## 8.   Electrical equipment

*Check*

8.1   That equipment which is not required for a long period is switched off
8.2   That the best tariff is negotiated with the Electricity Board
8.3   That the power factor control is adequate
8.4   That the intermittent use of electrical equipment is examined, with the object of lowering the maximum demand
8.5   That electric motors are matched to their required duties (eg, a 20 hp motor on an actual load of 8 hp uses 27% more power than a 10 hp motor)
8.6   That use is made of motors of the highest possible speed (eg, a 720 rpm, 10 hp motor on an actual load of 5 hp uses 17% more power than a 2,930 rpm 10 hp motor)
8.7   The feasibility of employing higher voltage supply to reduce transmission loss
8.8   That the contact points of all switchgear are regularly inspected and cleaned.

## 9.   Lighting

*Check*

9.1   That the best use is made of daylight by keeping windows and roof lights clean
9.2   That the layout of workplaces is such that those needing high levels of lighting are near windows or roof light
9.3   That lamps and fittings are maintained clean

9.4     That old lamps are replaced when their efficiency drops –
        this relates particularly to fluorescents
9.5     That dark backgrounds are avoided; paint furniture and
        decorations in light colour to give maximum reflection of
        light
9.6     That lights are switched off when not in use or not needed
        (due to natural light being good); consider photocell auto-
        matic control in large areas
9.7     That artificial lighting in areas well lit by natural light is
        separately switched from lighting in darker areas
9.8     That the most efficient type of lamps are employed
9.9     That the level of artificial lighting is not higher than neces-
        sary
9.10    That yard lights are not switched on in daylight hours or
        unnecessarily after dark
9.11    Someone is made responsible for turning lights off after
        staff have gone home and particularly after the cleaners.

## 10.   Records

*Check*
10.1    That comprehensive records are maintained of energy pur-
        chases and used to pinpoint unwarranted changes which
        might point to wastages having developed
10.2    That degree-day records are kept and used to check on
        space heating expenses
10.3    That records are kept of all essential maintenance of energy
        equipment
10.4    That a responsible person has been appointed to supervise
        re-lamping and regular cleaning of the light fittings
10.5    That a record is maintained of all suggestions aimed at
        energy conservation, and of consequent follow-up.

The above list cannot possibly be all-embracing. It must be viewed,
modified and amplified to suit particular situations.

Having achieved a good standard of house-keeping, one can
then progress to the consideration of a solar heating system to
further improve the comprehensive energy usage.

## EXAMINATION OF POSSIBLE IMPROVEMENTS TO EXISTING SYSTEMS

In seeking fuel savings, one must always examine the alternatives to establish the 'best buy', particularly when only limited finance is available. The following section illustrates this important principle.

### Example – hot water cylinder heat loss

A 3 kW electric immersion heater is operated continuously to heat water in a storage cylinder of 140 litres (31 gall) capacity.

The hot water cylinder is poorly insulated and thereby loses say 0.25 kW/hr.

Given efficient lagging, these losses may be reduced to 0.05 kW/hr.

Assume solar heating season is 8 months = 243 days or 5,832 hrs. Then possible heat savings during period = 5,832×0.2 kW = 1,166.4 kW, which at an electricity tariff of 3.125p/unit costs £36.45.

If money is invested in a solar system, then the anticipated saving would be similar, ie, £30 to £40 for a 3-person household.

The difference is, of course, that it may cost no more than £10 to insulate the cylinder efficiently, whereas the householder may pay over £500 to obtain an equivalent saving by solar energy.

A small investment in a time clock connected to switch the electric immersion heater so that hot water is available for 8 hrs/day instead of 24 hrs/day would reduce the overall heat loss and double the potential saving.

*Best buy: cylinder insulation and time clock*

### Example – roof insulation

The 'U' value for an uninsulated roof is, say, 0.45; for insulated roof, say, 0.25.

A 93 m² (1,000 ft²) dwelling is heated for, say 32 weeks each year, the bedroom floor being maintained at 15.6°C (60°F) for say 8 hrs/day. Then, the achievable saving is (6,451,200 Btu) 1,890 kW, which at an electricity tariff of 3.125p/unit costs £59.

Although it may cost about £50 to £60 to insulate the roof space efficiently, this is a much better proposition than a solar assisted hot water supply.

*Best buy: thermal roof insulation*

The relative benefits of providing double glazing and/or cavity wall insulation should also be considered. A decision must depend on the location of the building and on the local fuel costs. Clearly, a better case exists for double glazing/cavity wall insulation where external temperatures are low, wind exposure severe and fuel costs high.

# 9

# MAINTENANCE

'Install and forget' is an advertising slogan not infrequently applied to solar panels; the reality seldom bears out such a boast! The selected anticipated life expectancy of solar collectors may well be twenty-five trouble-free years, but it is most unlikely that any solar collector or array of collectors may be left unattended for more than about five years without impairing its collecting efficiency. Indeed, even more frequent attention may prove of advantage.

Many solar systems installed in the southern states of North America at the beginning of this century fell into disuse and brought disrepute to the idea of solar collection because of poor application and corrosion problems, which brought with them the need for complete replacement rather than repair. Quite a number of these early installations were provided without sufficient knowledge and attention to matters such as unprotected metal and plastic, deterioration, discolouration, atmospheric, chemical and radiation attack, the importance of correct fixings, weathering, corrosion and anti-freeze techniques.

Attention to the following maintenance check list is advised and will appreciably prolong the efficient operational life of collector systems:

1. Over a period of time, dust and dirt will collect on the surfaces which are exposed to wind and weather, as well as to possible dust-storms, sea-water spray, industrial pollutants, insects and bird droppings.

*Check:* That the surfaces are free from foreign matter and bodies. If not, scrape clean and hose down with water jet. Scrubbing and the use of detergents may be necessary in bad cases.

2. Absorbers are usually manufactured of copper, aluminium or steel with selective coatings. Truly selective coatings are prepared by the use of acid-bath, electrostatic processes, etc; most of these have a high integrity over a long period of time; others have not. Numerous paint makers are offering paints as selective coatings; such painted surfaces are not truly selective to solar energy reception. Paint on metal may be particularly suspect.

*Check:* Periodically check on the coatings. Stock a small quantity of suitable matching paint for touching up of worn areas of painted coatings. Clean defective area before applying the paint. Damage to selective (non-painted) surfaces cannot be so easily repaired – refer to installers.

3. Condensation within the surfaces of a collector can cause oxidisation, promote electrolytical attack between dissimilar metals and/or encourage the growth of algae.

*Check:* On presence of condensation, If detected, examine the installation and ascertain the cause. Rectify by cleaning, repainting as appropriate, and then repairing the fault. The services of a specialist may be required for an extensive or difficult repair.

4. Focusing and semi-focusing collectors have reflective surfaces which may become discoloured (brilliant white turns yellow, deep blue turns to light blue) by ultra-violet decay or by dirt, dust and grime.

*Check:* On condition of the reflective surfaces. Clean and recondition to reinstate as new. This involves re-polishing metal reflectors and repainting painted surfaces.

5. Roofs and fixing may suffer some movement due to settlement or climatic conditions, leading to gaps and/or defects in the weathering arrangements.

*Check:* On soundness of the weathering arrangements. If defects are found, plug or fill gaps with weather-proofing material. Major defects may require the skill of a builder. It is most important that the weathering aspect is checked at short intervals, say annually, to obviate creeping damage due to water penetration or dampness.

6. Collector fixings may suffer movement or distortion due to exposure to high temperatures, building settlement or inability of rigid fixings to accept thermal movement.

   *Check:* That fixings have remained sound. Excessive movement of fixings may cause consequential damage to pipes which are connected to the collector; also to the structure/ roof to which the fixings are attached.

7. High winds, snow and ice may cause damage to the collector installation.

   *Check:* After a period of severe weather conditions that the installation has remained sound. Repair defects as appropriate.

8. Not all sealants used for the frames of collectors are stable to ultra-violet radiations and some deterioration may occur. A defective seal will permit water and dampness penetration into the collector, leading to condensation and associated defects. Excessive hardening of the seals may occur.

   *Check:* The condition of the seals between the frame components. Excessive hardening will leave the seals in brittle condition. In bad cases, complete re-sealing will be necessary.

9. In direct-circulation water systems (ie, without a heat exchanger), there is much turnover of water. In hard water districts, severe scaling or furring-up of the pipes may occur. In soft water areas, corrosion is more likely.

   *Check:* Periodically on the condition of the pipes and repair/ renew/chemically clean as appropriate.

10. In hard water districts, the electric immersion heater may suffer failure due to hard scale forming on the immersed element.

    *Check:* Periodically at the commencement of the main solar heating season, that the electric immersion heater is operational.

11. In installations which use anti-freeze concentrate fluids, such as ethylene glycol, degradation into organic acids is possible,

with a consequent reduction of the pH of the solution. The fluid must be replaced before it has deteriorated to an extent where it no longer offers frost protection.

*Check:* Re-charge the anti-freeze on an annual basis (more or less as in a motor car). The installer must leave behind clear instructions as to the periodic re-charging of the anti-freeze liquid. The user must be particularly made aware of the toxic character (if any) of the liquid used to charge the system. The appearance of any pink, blue or green traces of glycol at the draw-off tap is likely to indicate a fracture or leak between the primary and secondary sides (circuits) of the system; urgent specialist attention is then required.

12. Header water tanks are generally controlled by ball (float) valves. A defective float will gradually fill with water, sink and will then cause continuous overflow. Similar nuisance will be caused by a worn washer on the ball valve.

*Check:* That float and ball valve function correctly. A defective float necessitates replacement of float by unscrewing the defective one and replacing with a similar new one. A defective ball valve generally requires re-washering; this is done by removal of the split pin from the valve assembly, unscrewing of the valve, removal of the worn washer and replacement of same.

13. Failure of hot water circulation between collector and heat store.

*Check:* In case of thermosyphon system, whether there are any air locks or physical blockages and clear same. In case of pumped system, check whether pump has remained operational. Check the electrical switches and fuses. For the majority of installations, canned-rotor (sealed rotor) pumps will be used. These cannot be repaired in situ if motor or impellor have become damaged. However, some pumps have a clutch; release this a number of times and turn pump by hand – this may clear a blockage. The defective pump is generally handed to the supplier in part-exchange for a new pump. In case of belt-driven pump, check belt tension and gland seals periodically.

14. A number of isolating and/or control valves are likely to be fitted.

Defective gland seals will lead to leaks and, in hard-water areas, to damaging scale deposits. In soft water areas, leaking valve seals will cause corrosion and discolouration.

*Check:* All valve seals periodically and re-pack as appropriate. Badly deteriorated valves should be entirely replaced.

15. Oil-filled systems must not lose fluid via leakages.

*Check:* Over the system. If leaks are present, repair them and top up the fluid in the system.

16. Air vent points are likely to be fitted. These may be automatic (rare in solar systems) or manual (common).

*Check:* That the air is periodically vented from the system. Failure to do so will impede the circulation of water within the installation.
Air vents tend to drip when not fully tightened after being vented. Such drips may cause corrosion and/or scaling. In case of such defects being found, clean the valve and venting needle, screw in firmly and check that there are no leaks when the valve is closed.

17. One, or a number, of drain cocks are likely to be provided.

*Check:* That drain cocks do not dribble; this may cause consequential damage. Repair or replace defective drain cock(s).

18. The aluminium parts of a collector may suffer corrosion, possibly due to defective anodising, careless handling during transport or installation and electrolytic action between copper and aluminium.

*Check:* For severe and progressive pitting of the aluminium. There is no cure; replacement will be required. Temporarily installed aluminium collectors may suffer internal corrosion, leading to leaks through the metal. Defective aluminium coatings in contact with another metal (particularly copper) under moist conditions are likely to suffer corrosion due to aggressive electrolytic action.

19. Copper pipes are quite commonly jointed with fusion jointing methods employing solder and flux, the use of flux and the presence of excess flux may cause corrosion.

*Check:* Low fusion silver solder is known generally to have very limited chemical reaction with copper; hence the widespread use of fusion type water fittings. However, it is known that in some areas (in the UK, notably in Essex), there have been a number of failures. The tell-tale blue/green coatings on the inside of the tube and the eventual pitting and failure lead to only one precaution following failure: the replacement of the fusion joints by compression type fittings.

20.   Water storage tanks and cylinders made of metal do not have an indefinite life. Failure of such components generally commences with small leakages which will eventually lead to a major water discharge causing severe damage through flooding.

*Check:* Condition of the cold water tanks, internally and externally. Observe the external condition of the hot water storage cylinder, particularly around the immersion heater connection and at bolted head fixings. Decide whether tank/cylinder replacements are necessary.

21.   Temperature and safety controls will be provided. These must be in prime condition to fulfil their function.

*Check:* The correct functioning of the controls and safety features. Simulate controlled conditions and observe control action. If electric controls are found inoperative, check on fuses before inspecting the equipment for faults.

22.   The thermal insulation covering to pipes and tanks/cylinders is likely to deteriorate over a period of time. Efficient thermal insulation should be maintained.

*Check:* That thermal insulation has remained in place and that jointing tapes between sections of insulation have not come adrift. Reinstate defective insulation and supports to same.

23.   Solar panels may be damaged by 'other trades' working on the roof, such as television aerial or roof repair specialists.

*Check:* That workmen engaged on the roof will take care/ have not damaged the collectors. Repair as necessary.

24.   Swimming pool plastic collectors must be ultra-violet and chlorine stable and the welding of plastic to plastic must be by well-tried and tested methods and materials.

*Check:* Rigidity of plastic sheets and headers. If the slightest looseness is apparent, there is a failure. The only recourse is to the manufacturers. Do not attempt D.I.Y. repairs.

25. Some manufacturers of collectors offer up to five years guarantee on their products, a lesser number up to ten years and only the rare or rash one will venture to offer fifteen years. Customers should not be beguiled by the offer of over-long guarantee periods – none can guarantee that the manufacturers will remain in business that long; hence, this factor should not sway a decision unduly. Experience has shown that the effective life of a collector and associated equipment can be greatly prolonged by intelligent and diligent maintenance.

Collectors are most commonly located in or on roofs and in other more, or less, inaccessible locations which will require ladders for access. An element of specialist knowledge may also be involved in aspects of maintenance. The owner of a system is therefore well advised to leave the routine maintenance to a contractor engaged in this particular field. It is prudent to seek and obtain an assurance from the installer that he will be willing and capable to undertake the subsequent routine maintenance and to obtain a quotation for the cost of this at the time of ordering the original installation.

26. Brass water pipe fittings are commonly used. Brass contains mainly copper and zinc. However, in some areas, the characteristics of the water cause dezincification (dissolving of the zinc, leaving the original brass porous and subject to leakages).

*Check:* Before installation whether brass fittings may be safely used in the locality of the installation (the local Water Company will generally provide the required advice). If brass cannot be used because of known dezincification difficulties, employ gun metal fittings instead. Replace dezincification damaged fittings with gun metal fittings.

27. The back-up heating to a swimming pool may be via a boiler, fired by solid fuel, oil or gas. In view of the presence of chlorine in the swimming pool water, the contaminated water circulating through the filter and swimming pool must not be circulated directly through the boiler, as rapid corrosion of the boiler will then take place. Special boilers with glass lined or otherwise coated internal surfaces are available;

whilst these will give a much better life expectancy than unlined boilers, they tend to be more bulky in size and any defect in the protective coating will upset the protective effect.

*Check:* In the event of boiler failure, whether this was caused by the direct circulation of swimming pool water through the boiler. Replace by an indirect system (boiler and heat exchanger). Some manufacturers offer a boiler for swimming pool use which incorporates a separate heat exchanger so that the swimming pool water does not pass through the boiler itself, although boiler and special heat exchanger are fitted into a common casing.

# 10
## PAYBACK

Enthusiasts may install solar energy systems regardless of economic viability, but the great majority of potential customers for such systems need to be convinced that tangible benefits are at hand. Mature consideration of the economic viability of a solar energy scheme should be based on an assessment of the 'payback period', at the end of which time-span the system should have repaid its initial cost out of the resultant savings. The factors to be evaluated include:

a) *Liabilities* The initial cost of the selected installation.
Interest charges on the capital, assuming that this has been loaned or would otherwise have been profitably invested.
Operating and maintenance costs.
Insurance.
Anticipated life of the installation before it has to be replaced by a new one.

b) *Tangible benefits* Amounts of solar energy to be collected (nett after accounting for heat losses).
Savings in energy cost arising out of the substitution of solar energy for conventional fuel (this may be oil, gas, coal or electricity, depending on circumstances).

c) *Uncertainties* Fluctuation in the rates of interest on borrowed/invested monies over the long time-span involved.
Average inflation rate over period.
Movement in prices of fossil fuels over period.

d) *Longer term forecasting* This may allow for an eventual reduction in the unit cost of solar energy collectors, as more experience is gained and volume production of collectors is achieved. Similarly, installation costs are likely to fall as more firms accumulate expertise in the integration of solar energy systems with buildings and architecture.

### PAYBACK – A SIMPLE APPROACH

The economic benefit of the utilised solar energy must be stated in terms of the cost of the alternative fuel which it displaces. By way of a simple example, a comparison is made below between a solar energy system and an electric immersion heater operating on the 'on-peak' electrical supply for the heating of domestic hot water. It is assumed that the electricity will be converted into heat at the point of use at 100% efficiency; whilst this is not strictly correct, allowing for heat losses from the surfaces of the storage vessel, similar losses would occur with both systems and hence cancel each other out in so far as the comparison is concerned.

Arguably, the selection of an 'on-peak' electric hot water system offers an unfair comparison, as it will show up the use of solar energy to much better advantage than the lower cost fuels, such as goil, gas or 'off-peak' electricity. However, it is useful to consider the maximum benefit obtainable from solar heat and then apply appropriate factors to other situations. Also, a great many domestic hot water installations rely on the use of electric immersion heaters, either throughout the year or only during the summer.

### Example

It is assumed that a solar energy installation will be provided by a competent contractor; overall, the cost therefore includes overhead charges and profit.

| | |
|---|---|
| Cost of installation, including mechanical, electrical and associated builders' work: | £450 |
| Available solar energy: | 900 kWh/m² per annum |
| Collector area: | 4 m² |
| Assumed collector efficiency: | 45% |
| Average cost of electricity: | 2.0 pence/kWh |
| A circulating pump will be incorporated: | |

Amount of solar energy collected   $= 900 \times 4 \times 0.45$
$= 1{,}620$ kWh/annum

Saving of electrical energy:       $1{,}620 \times 0.02$
$= £32.40$/annum

Deduct electricity consumption
of circulating pump, say:          £5.00
Net value of saving:               £27.40/annum
Payback period =

$$\frac{\text{initial capital cost}}{\text{saving/annum}} = \frac{£450}{27.40} = 16.4, \text{ say } 16 \text{ years}$$

## Inflation

The above computation assumes a zero inflation rate and makes no allowance for an increase in fuel costs over the life-span of the installation.

In practice, the actual payback period will be materially changed by variations in the original capital cost and in the cost of the alternative energy prices. By way of illustration, one can vis-ualise someone installing a system on a 'do-it-yourself' basis, in which case labour charges, overheads and profits would not arise, so that one might talk of a capital cost of only £250. Clearly, this would greatly improve the payback time. As regards fuels, the

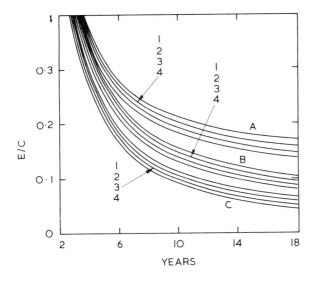

10.1 Analysis of payback time

forecast is for a continuing escalation in the price of fossil fuels and electricity, so that some allowance for the inflationary factor must be included within the payback calculations in order to obtain a fairer comparison.

Tables 10.1 and 10.2 illustrate the effect of inflation on the payback periods.

**Table 10.1  Effect of increasing electricity prices (at 10% relative/annum) on annual savings, starting with an annual saving of £31 in 1975 (real prices)**

| Year | | Saving £ Per annum | £ Cumulative |
|------|---|------|------|
| 1975 | 1 | 31 | 31 |
| 1976 | 2 | 34 | 65 |
| 1977 | 3 | 38 | 103 |
| 1978 | 4 | 42 | 145 |
| 1979 | 5 | 46 | 191 |
| 1980 | 6 | 51 | 242 |
| 1981 | 7 | 56 | 298 |
| 1982 | 8 | 61 | 359 |
| 1983 | 9 | 67 | 426 |
| 1984 | 10 | 74 | 500 |

**Table 10.2  Approximate payback periods at different fuel and capital costs at constant fuel prices for solar collector of 4 m² collecting surface**

| Fuel costs p/kWh | October 1975 | Payback period in years for different capital costs | | | |
|------|------|------|------|------|------|
| | | £250 | £300 | £400 | £500 |
| 0.5 | | 28 | 75 | 100 | 125 |
| (0.82 | Natural Gas) | | | | |
| 1.0 | Electricity off-peak | 14 | 23 | 31 | 39 |
| 1.5 | | 9 | 14 | 18 | 23 |
| 2.0 | Electricity on-peak | 7 | 10 | 13 | 16 |
| 2.5 | | 6 | 7 | 10 | 13 |
| 3.0 | | 5 | 7 | 8 | 10 |

Assuming a 10% relative annual increase in *real terms* in the electric 'on-peak' tariff, the payback period for the investment of £450 will reduce to only 9 years, instead of the 16 years projected in the example. With the 'do-it-yourself' installation costing £250, the payback period would be reduced to a mere 6 years, making this an attractive proposition.

### PAYBACK – A MORE SOPHISTICATED APPROACH

It is common practice to conduct an analysis for a 'return on investment' on a 'present value' approach, in which the initial investment (cost), annual outgoings and annual savings are discounted to their present value; the sum of the first two is then compared to the latter. A major component of such analysis relates to the assessed life-span of the equipment, which is fairly well known for conventional systems on the basis of past experience.

With respect to solar energy installations, one is operating with a developing technology, lacking in reliable feedback concerning the life expectancy; an error in the estimation of this factor could produce misleading results. To allow for this aspect, the 'payback-time' analysis is best used, as it draws particular attention to the importance of the life-span of the system when analysing the results of the economic assessment.

The payback-time is defined as the time at which first cost and annual expenses with compounded interest equal the total savings of energy cost with compounded interest.

At the end of the payback-time the *value of the investment, C*, is

$$A = C \, (1+r)^n$$

in which $r$ is the annual interest rate. The annual recurring expenses (such as insurance, replacement and maintenance), referred to as the annual fixed charge rate $mC$, accumulates (as annuity) to

$$B = mC \, \frac{(1+r)^n - 1}{r}$$

The *income* from such a system, the accumulated annuity of the annually displaced conventional energy $E$, is

$$D = E \, \frac{(1+r)^n - 1}{r}$$

From the basic condition for the payback-time

$$A + B = D$$

one derives an expression for the *ratio* of the annually displaced energy cost to the first cost as

$$\frac{E}{C} = m + \frac{r}{1 - (1+r)^{-n}}$$

This ratio approaches $(m+r)$ for large values of $n$ and is substantially larger than $(m+r)$ for less than 20 years.

Clearly, the higher the interest rate, the longer the payback-time, as it takes longer to compensate for the more rapid appreciation of the first invested capital via the annually saved energy cost.

From Fig. 10.1, family A, one sees that for reasonable interest rates and a payback-time of about 10 years, a ratio of $E/C$ of about 0.2 is required.

The $E/C$ ratio for most solar energy systems is usually less than 20%, hence payback times substantially in excess of 10 years arise on this basis, which ignores inflation.

### Inflation

The payback time decreases materially when inflation is considered. Thus, so long as the inflationary increases in the alternative fuels outstrip the recurrent costs, when the effect of inflation on conventional fuels is considered, the following modified relationship is established:

$$D_I = \frac{(1+r)^n - (1+e)^n}{(r-e)} \ E$$

in which the constant inflation rate for energy is $e$ and the savings in year $x$ are

$$E(1+e)^x$$

In $x$ years this would yield savings of

$$E(1+e)^x (1+r)^{n-x}$$

The annual recurrent expenses would inflate at a different annual rate, so that the equation for this, given above, will be modified to

$$B_1 = \frac{(1+r)^2 - (1-i)}{(r-i)} \ mC$$

in which $i$ is the appropriate inflation rate. When $e$ is larger than $i$, the modified expression $A+B_1=D_1$ will give a much reduced $E/C$ ratio:

$$\frac{E}{C}=a\,(r-e)+b\,\frac{r-e}{r-i}\,m$$

in which $a$ equals

$$\frac{1}{1-\left(\dfrac{1+e}{1+r}\right)^n}$$

and $b$ equals

$$\frac{1-\left(\dfrac{1+i}{1+r}\right)^n}{1-\left(\dfrac{1+e}{1+r}\right)^n}$$

The ratio $E/C$, for large value of $n$, approaches

$$\frac{E}{C}=r+\frac{(r-e)}{(r-i)}\times(m-e)$$

On the assumption of an inflation rate of 9%, one can plot the curve of Family B (Fig. 10.1); this indicates the resulting decrease in the ratio $E/C$ or in $n$. The curves for families A and B assume initial yearly expenses to be at 5% of the overall cost of the installation.

Based on present information, the life expectancy of a conventional solar heating installation in a temperate environment is about 20 years, after which time it has to be replaced. Hence, a payback-time of less than 20 years would make a worthwhile investment, though much shorter payback-times are likely in the usual circumstances.

When the payback-time period has been reached, with the system still intact and working, the solar energy system will continue to deliver energy at the recurrent costs level only, ie, at a profit.

*References:*
K. W. Boer, 'Payback of Solar Systems', *Solar Energy,* vol. 20. page 225, Pergamon Press, 1978

*Solar Energy: a UK Assessment,* page 28, UK Section of the International Solar Energy Society, May 1976

# 11

# LEGISLATION AND CODES OF PRACTICE

The installation of a solar heat collector system involves a number of aspects which may impinge upon a shared environment and common services. The selection and assembly of the materials being used may be subject to Standard Codes of Practice and/or government or local authority regulations. The following check list is relevant:

## 1. Siting of external system components

*Check:* Requirements for planning approvals/consents from neighbours.

## 2. Installation

*Check:* Associated building and civil engineering work to conform with all relevant building bye-laws. (For the UK see Building Regulations 1976.)

## 3. Safety

*Check:* Access to equipment for repairs and maintenance to conform with relevant safety standards. (For the UK see the Health and Safety at Work, etc, Act 1974.)

## 4. Standards of materials

*Check:* Specified components and materials to conform to relevant Codes and Standards. (For UK consult the British Standards Institution.)

## 5.  Water supply

*Check:*     Availability of clean water supply; installation and use of water to comply with restrictions and regulations of the water utility company, which will be concerned to avoid contamination and waste of water. (For UK see British Standard Code of Practice CP 310: 1965 – Water supply.)

## 6.  Electrical supply

*Check:*     Availability of adequate electric supply for the circulating pump and back-up electric heaters, as may be appropriate and compliance with electrical codes and safety regulations. (For UK see Regulations for the Electrical Equipment of Buildings issue of 1976, published by the Institution of Electrical Engineers.)

## 7.  Drainage

*Check:*     Availability of gully, or run-off for emptying system and compliance with water utility company regulations.

*Check:*     Avoidance of nuisance to adjoining properties or highway.

## 8.  Anti-freeze solution

*Check:*     Suitability for prevailing temperature conditions at site.

*Check:*     Whether leakage of toxic or corrosive anti-freeze solution could cause damage to installation and/or surrounding environment/persons.

## 9.  Heat transfer fluid

*Check:*     Suitability for prevailing conditions of operation and likelihood of frost damage.

Whether use of a toxic or corrosive fluid might lead to early deterioration of equipment. If toxic/corrosive fluid *must* be used, take precautions to avoid damage in the event of a leakage of the fluid.

Whilst the above check list might appear all-inclusive, it is not. More legislative decisions and guidance are required to form an all-embracing legislative package for solar heating installations, for instance:

Direction to ensure that fossil fuels must not be *exclusively* used for the heating of swimming pools in areas where solar heat is available. As the pressure on fuel resources increases, similar directions may be required concerning the heating of domestic hot water supply.

Codes of Practice relating to the materials being used for solar panels, with the specific exclusion of certain materials and finishes for reasons of public safety (eg, certain forms of mirror-like finishes, currently sold or proposed with solar panels, reflect sunshine and such reflected glare from single collector installations, or groups of same, could cause annoyance and possibly dangerous conditions to motorists and aircraft). Relevant Codes of Practice are required urgently and before a serious related accident occurs.

At the present time, there is no assurance that the owner of a solar collector will indefinitely continue to enjoy the benefit of the solar collection, as this may be legally disrupted or aborted by someone placing an obstructing building in the path of the solar radiation. Legislation is required to clarify rights to the uninterrupted receipt of solar radiations. (The UK already has legislation relating to 'ancient lights', giving the legal right to continuous receipt of light through the window. Similar or adapted, legislation could be provided to deal with solar radiation.)

Codes of Practice concerning water supply require to be enlarged to include specific regulations, requirements and restrictions relating to the use of the various solar collection systems with reference to the periodic emptying of systems to avoid frost damage.

A Code of Practice may well be useful to recommend appropriate treatment of water in solar collection systems.

The rapid rise in the number of different solar collectors being offered for sale calls for the urgent preparation and issue of an agreed Test Code for such collectors. At the present time, the purchaser has to evaluate the relative claims made for different constructions, coatings, etc, without formal guidance.

A Code of Practice is required to guide users of anti-freeze and heat transfer fluids and solutions, with particular reference to the use of toxic and/or corrosive liquids. Some advertised solar energy systems employ so-called 'improved heat transfer fluids', such as certain oils and ethylene glycol. Fracture of a system component containing or conveying such fluid could result in leakage into the water, with the possibility of someone then drinking the contaminated water. Leakage of corrosive fluids could cause a host of difficulties.

In the USA, the Energy Research and Development Administration (ERDA) monitors the progress of research and development and the passage of legislation relating to energy. Certain legislation has already been enacted and various standards have been imposed on manufacturers and installers of solar energy equipment. Many other countries have so far failed to show a sense of urgency in this matter.

It is likely that, in the not too distant future, legislation will be introduced in the UK to regulate and guide alternative energy programmes in the interest of fuel conservation. It will then become the clear duty of the authorities to establish standards of safety to protect the public at large from misuse or malpractices which are likely to arise in the field of alternative energy usage.

# 12

# THE FUTURE OF SOLAR ENERGY HEATING

Research and development into the application of solar energy to heating applications is growing rapidly, as governments feed ever-greater sums of money into solar energy-related programmes. Major advances in materials and techniques are likely, making forecasting a hazardous task. However, the following trends are discernible:

Improvements in the basic concept and design of solar collectors to upgrade the collection efficiency, especially in cloudy conditions, to enhance cost effectiveness and economic viability relative to other methods of heat generation

Design and construction of building architecture to facilitate and co-ordinate solar heat collection

Replacement of the conventional water storage system by 'change-of-state' heat stores to reduce the space requirement for the heat store and to increase its storage capacity

Substitution of more durable, lighter-weight corrosion-free materials, particularly plastics

The solar heat-storage will come into widespread use as a generator of high temperature water and steam for heating and air conditioning, see Fig. 12.1

Combined arrangements of solar collectors and heat pumps to achieve higher storage temperatures and/or to provide air conditioning, see Fig. 12.2

Extension of solar heat usage to space heating, particularly by circulation of warm air, refrigeration, cooking and air conditioning, see Fig. 12.3

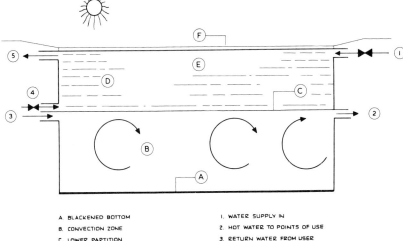

A. BLACKENED BOTTOM
B. CONVECTION ZONE
C. LOWER PARTITION
D. INSULATING LAYER
E. UPPER PARTITION
F. THIN FRESH WATER COVER

1. WATER SUPPLY IN
2. HOT WATER TO POINTS OF USE
3. RETURN WATER FROM USER
4. CONCENTRATED SALT WATER
   SUPPLY IN
5. SALINE WATER RUN · OFF

12.1 Solar pond diagrammatic arrangement

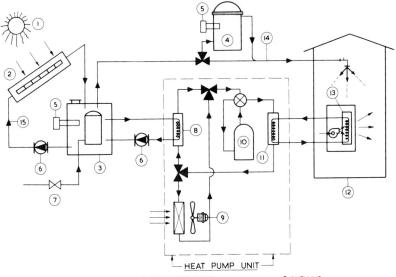

1. SUN
2. SOLAR COLLECTOR
3. WATER STORAGE TANK
4. HOT WATER TANK
5. IMMERSION HEATER

6. CIRCULATING PUMP
7. COLD WATER SUPPLY
8. AIR SOURCE EVAPORATOR
9. WATER SOURCE EVAPORATOR
10. COMPRESSOR

12. DWELLING
13. FAN ASSISTED TERMINAL HEATER/COOLER
14. HOT WATER TO TAPS
15. CIRCULATION BETWEEN COLLECTOR
    AND HEAT SPACE

12.2 Solar collector and heat pump

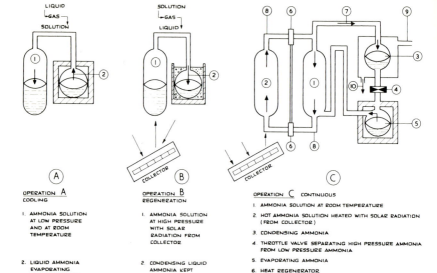

12.3 Solar collector and refrigerator

Acceptance of the solar heat contribution as essential for the heating of outdoor and indoor swimming pools

Use of solar heat for the heating of green houses, employing 'wet-earth' and similar low cost heat stores

Widespread adoption of solar heating systems in agriculture, forestry and industry for the drying of crops, timber and chemicals

Looking far beyond the present into the next century, it is probable that solar collectors orbiting in space will focus and beam copious solar energy to receivers on earth for the purposes of major heat and power generation

The issue of legislation and byelaws restricting the use of fossil fuels for certain applications in favour of solar energy and regulating the applications to buildings of solar energy installations.

As the first full scale orbiting solar collector beams the sun's bounty to the earth, mankind will cross the threshold into an age of golden opportunity.

# APPENDIX:
# CONVERSION FACTORS

Systeme International (SI) units are primary units used in this book with conversion, wherever possible, to imperial units. The following symbols, conversion factors and definitions are given to assist in the understanding of both unit methods. Certain units commonly used in the USA and elsewhere are also included.

## MULTIPLES OF SI UNITS

| Factor | Name | Symbol |
|--------|------|--------|
| $10^9$ | GIGA | G |
| $10^6$ | MEGA | M |
| $10^3$ | KILO | k |
| $10^{-2}$ | CENTI | c |
| $10^{-3}$ | MILLI | m |
| $10^{-6}$ | MICRO | $\mu$ |
| $10^{-9}$ | NANO | n |

## LENGTH

| | |
|--|--|
| 1 millimetre (mm) | 0.039 inch (in) |
| 1 metre (m) | 3.281 ft or 1.094 yard (yd) |
| 1 yd | 0.914 m |
| 1 kilometre (km) | 0.621 miles |
| 1 mile | 1.609 km |

## AREA

| | |
|---|---|
| 1 square centimetre (cm²) | 0.155 in² |
| 1 square metre (m²) | 10.764 ft² or 1.196 yd² |
| 1 square inch (in²) | 6.452 cm² |
| 1 square ft (ft²) | 0.093 m² |
| 1 square yd (yd²) | 0.836 m² |
| 1 square kilometre (km²) | 0.386 mile² |
| 1 square mile | 2.590 km² |

## VOLUME

| | |
|---|---|
| 1 cubic centimetre (cm³) | 0.061 in³ |
| 1 litre | 1,000 cm³ or 0.220 gallons |
| | or 0.264 gallons (US) |
| 1 gallon | 4.546 litre |
| 1 gallon (US) | 3.785 litre |
| 1 barrel (oil) = 35 gallons | |
| or 42 gallons (US) | 0.159 m³ |

## MASS

1 kilogramme (kg) = 1000 g  =  2.205 lb
1 tonne (metric)   = 1000 kg=  0.984 ton

## PRESSURE

1 N (Newton) m² = $1.020 \times 10^{-5}$ kg/cm² = $1.450 \times 10^{-4}$ lb/in²
1 kg/cm² = $9.808 \times 10^4$ N/m² = 14.223 lb/in² = 0.968 atmospheres
1 lb/in² = $6.895 \times 10^3$ N/m² = $7.031 \times 10^{-2}$ kg/cm² = $6.805 \times 10^{-5}$
atmospheres
1 atmosphere = $1 \times 10^5$ Pa (Pascal) = 760 mm Hg = 29.92 in Hg = 33.90
ft water = 1.013 N/m² = 1.033 kg/cm² = 14.696 lb/in²
1 bar = $10^5$ N/m² = 0.987 atmospheres

## DENSITY

| | | |
|---|---|---|
| 1 kg/m³ | = 6.243 lb/ft³ | = $3613 \times 10^{-5}$ lb/in³ |
| 1 lb/ft³ | = $5.787 \times 10^{-4}$ lb/in³ | = 16.019 kg/m³ |

## TEMPERATURE

0° Centigrade   = 32° Fahrenheit (freezing point of water)
100° Centigrade = 212° Fahrenheit (boiling point of water at

normal atmospheric pressure)
To convert Centigrade to Fahrenheit multiply by 9/5 and add 32
To convert Fahrenheit to Centigrade subtract 32 and multiply
by 5/9

## HEAT, ENERGY AND POWER

The Calorie is the amount of heat required to raise 1 gramme of
water through 1° Celsius (Centigrade)
The British Thermal Unit (Btu) is the amount of heat required to
raise 1 lb of water through 1° Fahrenheit
Heat is a form of energy and the Joule (J) is commonly used as
a mechanical unit of heat
The fundamental unit of power is the Watt (W)

| | | |
|---|---|---|
| 1 Btu | $= 1.055 \times 10^3$ joule (J) | $= 778.169$ ft lb |
| 1 calorie (cal) | $= 4.187$ J | |
| 1 kilocal | $= 3.968$ Btu | |
| 1 Watt | $= 1$ joule per second | $= 0.00134$ horse power |
| 1 kilowatt hour (kWh) | $= 3.600 \times 10^6$ J | $= 3.600$ MJ |
| | | $= 3.412 \times 10^3$ Btu |
| 1 Btu/h | $= 0.293$ W | |
| 1 kilocaloric/m² | $= 0.369$ Btu/ft² | $= 1.163$ W/m² |
| 1 W/m² | $= 3.6$ kJ/m²/h | $= 0.317$ Btu/ft²/h |
| 1 Btu/h/ft² °F | $= 5.678$ W/m² °C | |

The Langley is a unit of energy frequently used in the USA and
is equivalent to 1 calorie/cm²

## Energy equivalent of fuels

| | |
|---|---|
| Oil: | 1 barrel $= 6.3 \times 10^9$ J $= 5.7 \times 10^6$ Btu |
| | 1 tonne $= 4.47 \times 10^{10}$ J |
| | 1 Mtoe (million tonnes of oil equivalent) |
| | $= 4.47 \times 10^{16}$ J |
| | $= 1.24 \times 10^{10}$ kWh $= 4.24 \times 10^{13}$ Btu |
| Gasoline: | 1 litre $= 3.5 \times 10^7$ J |
| Natural gas: | 1 cu. ft $= 10^3$ Btu $= 1.05 \times 10^6$ J |
| | 1 m³ $= 3.7 \times 10^7$ J |
| Black coal | 1 ton $= 2.8 \times 10^{10}$ J $= 2.6 \times 10^7$ Btu |
| | 1 kg $= 3.1 \times 10^7$ J |
| | 1 lb $= 1.3 \times 10^4$ Btu |

Mtce (million tonnes of coal (black) equivalent)
$= 2.85 \times 10^{16}$ J $= 7.92 \times 10^9$ kWh
$= 2.70 \times 10^{13}$ Btu

Uranium metal:   1 ton $= 8.1 \times 10^{16}$ J (fully burnt)

# INDEX

*Figures in italic refer to illustrations*

Heat store
– building structure as 91
– definition 15
– energy input, relation to 94
– greenhouse application 100
– heat-of-fusion 91–94
– hot water systems 68, 69
– main storage methods 90–92
– materials, suitable 90
– open top 76
– pebble bed *31*, 91
– pump, effect of use of 78, 79
– relationship to associated system 95
– space heating 90–95
– water storage tanks 90, 91
Heat transfer fluid
– check on suitability 133
– definition 16
Hot water services
– cylinder heat loss 115
– design parameters 68
– direct supply system
  – corrosion 62, 69–73
  – diagram *61*
  – disadvantages 62
  – generally 61
– domestic supply 60, 61
– dump tank, direct system with *71*, 71, 72
– energy conservation 112
– flow rate and collector efficiency *82*
– flow rate and pressure loss *83*
– frost, precautions against 63, 67, 68, 72
– generally 60
– gravity systems, disadvantages 75
– heat store, system relation to 69
– indirect supply system
  – advantages of 67
  – circuits for *64*, 65
  – solar collectors imposed upon 62, 63, *64*, 65
  – storage cylinders, types of *63*
– low silhouette system *72*

– multi-collector multi-cylinder system 64, 65, *66*
– multiple collectors with single cylinder 62, 63, *65*
– pumped systems 78–83
– scaling and corrosion 69–75
– school ablution systems 60
– solar collector added to indirect system 62, *64*
– storage cylinders with heat exchange coils 62, *63*
– temperature levels, normal 60, 61
– thermosyphonic systems
  – generally 75–78
  – recommended pipe sizes 77
– water treatment 75, 76
Hour angle, definition 16

Immersion heater, maintenance 119
Inclined surface
– definition 16
– diagram *17*
Industrial processes 100, 101
Infra-red, definition 16
Insolation
– definition 16
– levels, determination of 45
– winter periods 56
Installation checks 132
Insulation
– definition 16
– double glazing and cavity walls 116
– flat plate collectors 26
– materials, use of 26
– roof, of 115
– thermal, maintenance 122
– thermal, provision of *61*, *63*, *64*, 66
Irradiance
– definition 17
– values, determination of 45

Latitude, definition 17

south essex college
FURTHER & HIGHER EDUCATION
THURROCK CAMPUS